To Bill

With appreciation for
your dedication to
the public interest
With best regards
Ted

December 1983

A DESIGN FOR FREEDOM

A Social Analysis of the Public Utility Process

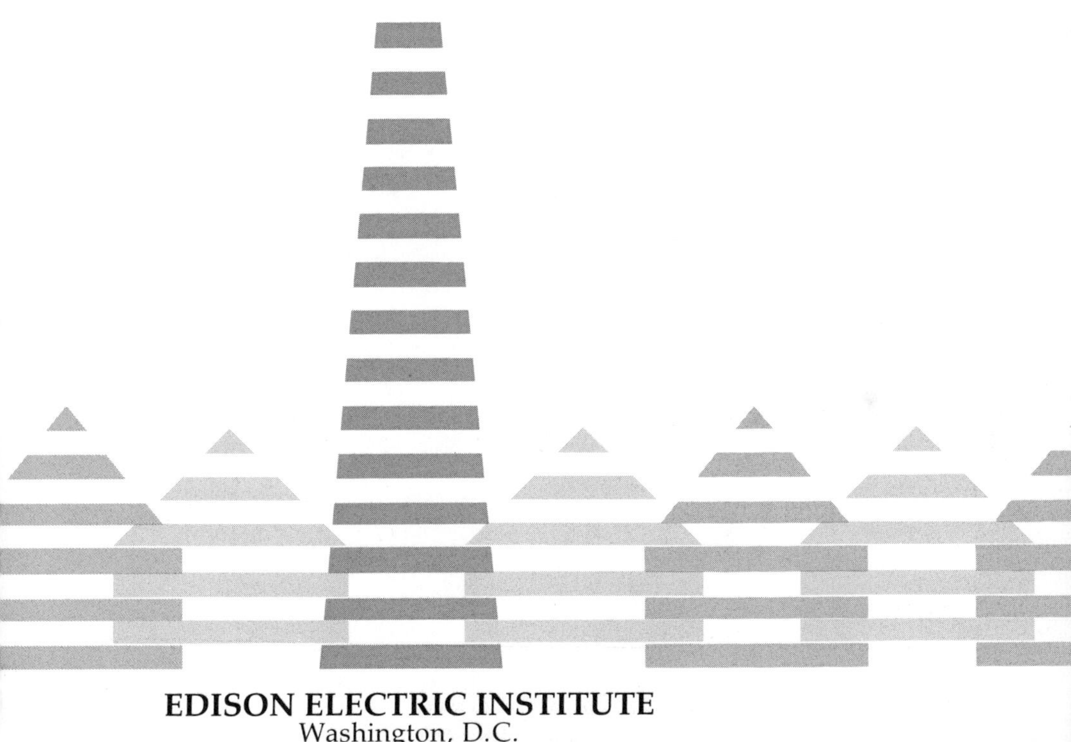

EDISON ELECTRIC INSTITUTE
Washington, D.C.

PREFACE

An electric utility company is a unique social organization. It brings together, within the American process and tradition, people and resources to provide an energy form that munificently contributes to our material well-being and prosperity. The fact is, the value of electric service is so taken for granted that it is accepted with little recognition of its significance or the complexity of the process necessary to supply it. The utter necessity of electricity is forgotten until a blackout or brownout. Then our dependence upon it becomes a stark reality.

The technological contribution, however, is only part of the story of a utility company. The other part—and perhaps the most important—is that it represents a social process that reflects the fundamental concepts of a free society.

The phrase "free society" is used glibly, but the actuality of a "free society" is perhaps our most precious possession, the total consequence of the American genius. Basically, it means a society where individuals can come and go as they wish, say what they think, do what they want, all without the constraint of social institutions dictating to individuals and groups and demanding their conformance. A living example of repressive restraint on groups and individuals is supplied to us by the events in Poland. In contrast, a free society lives on pluralism where many institutions can contribute to the general good without anyone becoming predominant and controlling.

In this kind of society, a utility company is a legitimate expression of the democratic process. It is basically an instrument of free enterprise and ingenuity, but it tempers enterprise with a constant consideration for public interest. It does not represent a plan or a preconception; instead, it is a product of a long history of social experimentation.

The fundamental objective of a utility company is to serve the public interest, but it does so without becoming a governmental agency. It represents the unique American ideal of independent action; it embodies a rejection of the easy way out, of turning everything over to the government, so characteristic of the approach in other countries. Instead, it remains private but subject to intense public scrutiny. It is a laboratory for the interaction between the public sector and the private sector. It is an extremely sophisticated process, and those who are involved in the management of utility companies are skilled practitioners of the art of social interactions.

This book examines this process and deals with the utility company, not as a static thing; not as a study of economics, of regulation, of law, or of technology; but as a living, moving flow of people and resource activity—of interaction and interdependence.

Above all, it is an endeavor to see some of the dimensions of a free society, not in its simplistic form relating only to individuals, but in the complexities of large-scale modern technological institutions and organizations. Only in our ability to master such processes can we remain free. In these processes are the seeds of the design for freedom.

CONTENTS

WHERE THE ACTION IS

History deals chiefly with political events, with wars and revolutions—dramatic and cataclysmic events. Despite their notoriety, however, political upheavals do not, after all, change things very much. If one looks at the ways people live and do things rather than at the concepts of their governments, it appears that other kinds of factors are more significant in determining how things are done.

Considered in the long view, even the French Revolution, cataclysmic as it was, did not change the social customs of the French people; and even in communistic Russian society, we see many factors characteristic of the days of the tsars. Our own American Revolution left American economic and social life not vastly different from pre-Revolutionary times, and perhaps it was only our Civil War, with its great impact on social customs, that really changed things.

This is not to say that governmental techniques, people in high places, and legal structures do not change, but in day-to-day life, on the functional level, many things remain the same. The French ate croissants both before and after their Revolution, and the Americans drank tea both before—with a short interlude—and after theirs. Daily habits, mores, customs, and procedures provide the significance of life and, in them, meanings reside and emotions are determined.

Because we like to deal in concepts and principles which can be reduced to logical packages, it is often difficult to think of issues in terms of their practical consequences. But this is the level of basic meaning.

In all areas of life, however, there is a way of looking at things which focuses upon the point of human contact. Thus, in the area of law, an eminent professor at Harvard University, Paul Freund, offers that our emphasis on the Supreme Court and its decisions is too dominating. In an article entitled "Two Cheers for the Court," he stated:

> "So I would like to get away from that emphasis on the Supreme Court. I don't think that in any event the most significant features of our legal system are reflected in the Supreme Court; they are reflected in the police courts, in the police stations, in the law offices, in legal aid offices, and so on."[1]

In short, the law touches us where we have a personal contact with it.

In the world of art, the meaning of music is the transcendental, inspirational, and visionary sense which the hearer experiences whether it comes from a sophisticated symphony or a simple song. Useful as it may be for several purposes, the accepted artistic judgment of professionals or experts as to the quality or merit of an art work is not the real criterion.

Nature and the vast outdoors exist in their own mysteries, for whatever they mean, but as they relate to human beings, they are meaningless unless, as William Wordsworth put it, the view of nature produces:

"... ; a sense sublime

Of something far more interfused,
Whose dwelling is the light of setting suns,

And the round ocean and the living air,

And the blue sky, and in the mind of man: ... "[2]

But nowhere is the functional impact better expressed than by William James, who in *The Moral Philosopher and the Moral Life* said:

"... the only force of appeal to us which either a living God or an abstract ideal order can wield, is found in the 'everlasting ruby vaults' of our own human hearts, as they happen to beat responsive and not irresponsive to the claim. So far as they do feel it when made by a living consciousness, it is life answering to life."[3]

The real forces in history, then, are those which relate to the way people act, think, and feel. In a dramatic, long-range view of civilization, the discovery of fire and the invention of the wheel are taken proverbially to be historical turning points. It is hardly disputable that the dominance of Western European civilization in the 18th and 19th centuries rested upon the inventions of (and subsequent monopoly in the use of) gunpowder and printing.

Think, then, in this kind of context about the impact of electricity. In all levels of life, primarily since the inventions of Edison, it has turned the world around. The changes brought about by the use of electricity are so commonplace that we take them for granted without a moment's thought or consideration, yet they have profoundly affected the lives of all of us.

Consider a few. The incandescent light literally turned darkness into day and vastly increased daily time for human living. The invention of the phonograph led to a phenomenal enlargement of those human experiences which are dependent upon hearing. The motion picture and television revolutionized entertainment and art experience, as well as making major changes in educational methods. The use of electrical machinery has made possible gigantic industrial developments, for example the Panama Canal with its giant locks and towing equipment operated by electric energy, a remarkable innovation in the use of electricity. Air conditioning, mostly electrical, radically changed living habits, making cities and working places more comfortable in the summertime. It made possible the burgeoning development of the Southwestern United States and brought new life to the Southeast. Without air conditioning, living in such places as Phoenix, Arizona, would be unbearable. Most recently, electricity is one of the factors that has made possible the potentialities of the space age.

Here, at the point of contact, where electricity is used and affects daily life—here is where the action is.

But some people argue that both the technical and social attributes of the industry which delivers that electricity are detrimental in total or in part. Questions arise whether the industry has really increased human fulfillment by creating the environment in which people can work all night, whether we have really improved human conditions by artificially cooling our cities and making the work year a full 12 months, without the traditional summer respite, whether the noise of sound equipment and the incessant nonparticipatory entertainment world of the films and television really has been beneficial.

While these value issues are now being discussed and debated, it is difficult to believe that society will forego the obvious pleasure, convenience, and respite from drudgery which electricity has provided. It is difficult to see how life would be more fulfilled by any such self-denial.

The life of the pioneers is romantic to read about, but the reality of crossing the Rocky Mountains by wagon train—or with pushcarts, as some Mormons did—is something else. Surely there must be few among us, including confirmed naturalists and environmentalists, who want to return to a pushcart civilization.

And in the competitive world, involving not only ourselves, but also the so-called second and third worlds, there may be no choice but to accept and appropriately mold to human purposes these marvels of our time.

It should also be recognized that these technological qualities did not just emerge. They were a product of a social system which encouraged and rewarded production and invention. Out of the free American system, the electrical world was produced. And while the ways to produce and distribute electricity are hardly as glamorous as the events of the invention of the incandescent light or the phonograph, it was the remarkable development of generating facilities—from the primitive Edison power plant to the mammoth generating facilities of today—and the accompanying development of transmission and distribution which made the commonplace use of electricity possible. And all of this—the unfolding of the incredible complex of pumps, boilers, pipes, telesystems, etc., which a modern generating plant represents—was a product of the private sector, the American corporation. Of course, utility systems have been developed abroad under governmental auspices, but clearly the leadership, at least in the formative years, came from the American system. Our utility systems unequivocally prove our country's leadership, notwithstanding some developments in other countries which threaten such leadership (notably in the field of nuclear power).

And so, the utility business, often thought of as prosaic and undramatic, teaches a lesson not only in imaginative technology, but also in the science of producing things for the general welfare and the common good. Today the utility industry is at the cutting edge of the issues of our time. It is vastly misunderstood, as are so many of the social and economic mechanisms of our society. And it is often criticized and complained about in the harshest of terms. But in spite of its critics, it has a gallant history, a present which contributes to the Constitutional objective of promoting the general welfare, and a future which is marked by a glowing promise.

What the following pages reflect is a bias, but it is one which is based upon experience and which can point to the results which have been obtained. It is logically possible that other social systems could have provided the electrical age, but it is a matter of history that it was developed by the private sector system. The record cannot be denied.

Nonetheless, the issues are so significant, so vital, so comprehensive that a bias is not sufficient. Instead, an objectivity, a search for deeper and better answers, must be pursued. Above all, a tolerance of viewpoint and an acceptance of goodwill on part of all who endeavor to understand and speak about the issues is a necessity, almost a constitutional or *a priori* requirement. An objectivity, a search for deeper and better answers, must be pursued.

This need was captured spontaneously, extemporaneously, casually, in an exchange a couple of decades ago between Robert Moses, who was conducting a hearing as Chairman of the New York State Power Authority, and a witness appearing before him. The witness, positive at the outset, and growing more vociferous as he proceeded, finally insisted that his approach was the American way. He was coolly interrupted by Mr. Moses who remarked: "Mr. _______, let's assume that we are all patriots and go on from there."[4]

More profound, perhaps, was the recollection of early English history and its application to a contemporary issue by Judge Learned Hand, the illustrious Chief Judge of the Court of Appeals of the Second Circuit. The Judge, who appeared in 1951 before the Senate Committee on Labor and Public Welfare, which was inquiring into the general issue of morals in public life, recalled a statement by Oliver Cromwell just before the Battle of Dunbar. Cromwell, in an endeavor to avoid what he knew would be a brutal conflict, wrote to the Scottish leaders: "I beseech ye in the bowels of Christ think that ye may be mistaken."[5]

Such a statement, said the jurist, with the second "ye" changed to "we" should be inscribed over the portals of every courthouse and legislative hall in the United States.

And so, with the recognition of competing views, the thoughts of this book are offered as a statement of the philosophy of the public utility process, and in the spirit of Cromwell's remark (both with and without the Judge's editing): ". . . think that we/ye might be mistaken."[6]

Chapter 2

THE UTILITY AS A RESPONSE TO ITS COMMUNITY

Most of the inquiries into the public utility industry start with the utility companies themselves. It would be better if they started with the community, for it is out of the community and its needs that the utility service arises. Thus, an examination of a utility company requires some basic understandings of the community itself.

The first inquiry should perhaps be a technological one: What need does the community seek to have satisfied by the utility? On the surface this seems totally simple. A community needs water; thus, a water company. A community needs gas, thus, a gas company. Likewise, to satisfy the community's need, a transportation company supplies transportation. And an electric company supplies electricity.

Perhaps this simplicity suffices for a water company and a transportation company, and perhaps even for a gas company. But when an electric company is considered, a simple approach becomes impossible. For it would not be wrong to say that electricity, technologically speaking at least, makes the modern community. The need which the community seeks to have satisfied by the electric company is not completely inherent in the community. The community's need for electricity changes as the community itself is modified and molded by electricity.

No one knew of electricity until its use became feasible. In all the centuries of human existence, only the current one has had electricity. Yet prior to the availability of electricity, society operated very satisfactorily. To a greater or lesser degree, society has always needed heat, but in the past that was supplied by shelter, clothing, and fire. Society has always needed light, but that was provided by daylight, fire, candles, or torches. Society has always needed to work, but that was furnished by muscle or animal power, or mechanical means such as the wheel and the lever. And it has always wanted entertainment, and that was supplied by "in person" performers.

But when electricity became available, the whole structure of human life changed. How it changed is most dramatically demonstrated by considering life at the turn of the 20th century and today. At the turn of the century there were no electric washing machines, dryers, refrigerators, freezers, air conditioners, radios, motion pictures, or television sets. The world was so vastly different from the one we live in that it is now difficult to comprehend it fully. Television, refrigerators, and air conditioners have literally accomplished a revolution in our way of life, not only directly in the way things are done, but also in many indirect ways. Air conditioning alone has changed work and recreation patterns, both daily and seasonally, and has created new communities in places hitherto thought to be climatologically inhospital. Inextricably, the life of the community is dependent upon electricity and the benefits which it bestows.

And so to inquire about the community's needs, vis-a-vis electricity, is to inquire into the needs of society which have been largely molded by the ingredient itself. And the entire question whether society demands electricity out of its inherent tendencies or whether its demands are stimulated and created by the opportunities presented by electricity use are irrelevant. The two—society and electricity—have become mutual functions of each other.

Within this framework, the specific technical needs of a local community are determined by its location, geography, topography, natural resources, and other physical characteristics. They are also determined by its economy, the nature of its industry, the composition of its population, its demographics, its social structure, and the organizations and associations, both public and private, which make up its cultural life. Affecting all of these factors is the magic of electricity and the utility company which delivers it.

The view is expressed that utilities have forced the use of electricity on the public, that the industry builds power plants to make money out of them. But the building of a power plant without a market for selling the output would lead to financial disaster, and no utility would build on such a speculation. The utility only builds when it sees the growth in load justifying the building of a new plant. It builds when its duty to serve indicates that it must expand.

The utility company, necessarily, is sensitive to these community needs, and analyzes and studies them in great depth. And the utility's technological program is inescapably designed to meet these needs.

Over and above the technical needs, the community reaches out and probes to find out how it should organize itself to provide for its economic and technological necessities. This reaching out is guided and molded by the national and cultural heritage, by the institutional patterns in which people live—the way they do things—and by the value systems which underlie both the national and local life.

The process of the community development determines the method by which its technological needs are to be satisfied. This mix of social and technological forces supplied the environment in which the utility company developed.

Obviously, in the all-embracing issue of how the community organizes itself to meet its technological needs, utilities are only a small, though far from insubstantial, part; but the way in which the community deals with the institutions relating to the utilities is symptomatic of how it will deal with other institutions.

Accordingly, it is necessary and decidedly not overpretentious to look at the mainstream of American life, in the beginnings, the present, and indeed even into the future, in order to understand the full scope of the utility process and the directions it may take. To perceive these foundations, there is a need to determine utility and energy policy and to separate and discern the factors which change from those which either do not change or change very slowly.

If we are going to really understand this, we have to understand the mainstream of American life, and that requires a look at how it was all formed—its beginnings, the present, and perhaps its future—in order to understand the whole scope of the utility process.

Among those considerations which do not change is our concern, or at least our expressed or alleged concern, for the individual. Perhaps the fundamental characteristic of the American milieu is the insistence on the inalienable supremacy of the individual.

The concerns for the rights of the individual are more sensitive than ever before, notwithstanding the tremendous metamorphosis which has transformed our society from rural life to an urban and industrialized civilization; the fact that society is mechanized, computerized, dehumanized, and influenced by the mass media in ways never previously thought possible; and the fact that our interests have become international and corporate in scope.

Freedom of the individual to follow his own wishes and desires, to be conventional or unconventional, appears to be greater than ever before. In our new lifestyle and expressions, we realistically and consciously relate our goals and objectives to the literature which grew out of the American Revolution—to the Declaration of Independence, to the Constitution, and to the Bill of Rights. This individualistic, pluralistic expression is no accident, but a consequence of the national heritage which infuses each community and its way of functioning.

Although so many years have gone by, not too much can be made of the fact that the United States was a pioneer country and still is, especially in comparison with other existing countries. While many countries today may be underdeveloped, only a few have faced the situation of beginning society anew in a virgin country with no indigenous traditions to bind them. The splendor which our forefathers encountered as they came to a vast, almost uninhabited continent, totally free from any established prescription for conduct or procedures, can hardly be imagined. The opportunity for personal creativity was unbounded; a rugged personal independence was inevitable.

The streams of thought which entered into forming the American community have been long debated and are not to be developed here. But the basic threads relevant to the argument are apparent and can be briefly mentioned. For a long time, the debt which America owed to European philosophic thought, the romanticism of Rousseau, and the logic of John Locke were considered to be the significant factors. Certainly they are not to be deemphasized, but more and more a simple functional approach becomes fruitful in aiding our understanding. People don't follow a plan or a philosophy; they are what they do, and the adaptations which people make in their daily lives, as they respond within a generalized framework of values to the problems and circumstances which confront them, are likely to be more significant than any system of organized thought.

The combination of customs, mores, social rules, concepts of law, and the demands of existence develop the conduct patterns which become institutions. Colonial America, which at least to some degree we gloss over as almost the American "Middle Ages"—slumbering between the mountain peak experience of discovery and initial colonization on one end and the Revolution on the other—was in truth a very vibrant society, constantly adapting to the immense opportunity inherent in it.

The period between 1607 and the Revolution represented a span of approximately 150 years, not substantially shorter than the life of the United States. During that time, lives were lived and communities were developed, which constituted the sum total of existence to those who were then living. It was no sleepy time waiting for our golden age to erupt. During this period, which in history is chronicled in terms of political events, lives and vibrant institutions were established. People lived by laws and the help of government, but their daily lives revolved around the educational, scientific, commercial, religious, and social institutions which they were developing. And these were vibrant and fertile institutions indeed.

The area occupied by the colonial people was characterized by some great similarities, the foremost being a uniform language. It was also characterized by the common experience of migration from Europe and conflict with the Indians. Besides the powerful similarities, the area was characterized by great diversities. People did what they wanted to do, what was the natural and commonsense thing to do amid the circumstances. Precedent and authority were of minor importance.

Thus, in the educational field, for example, almost from the moment the colonists landed, they began developing educational processes for young people which were natural and unrestrained by *a priori* professional requirements. Daniel J. Boorstin, in his book *The Americans, the Colonial Experience*, points out that by the time of the Revolution, there were at least nine degree-granting institutions in America. In England, because of a professorial monopoly, there were only two—Oxford and Cambridge—and they were deeply circumscribed by rights and privileges of the faculty. These professors would not acknowledge that anybody else who started a college really had professors. In other words, one had to be a Cambridge professor or an Oxford professor to be a true teacher. In contrast, the American university was a liberal, flexible instrument, responsive to the communities which created and controlled it, not to a self-perpetuating faculty.[1]

Out of this vigorous period came the American Revolution and the literature of the time, including the Declaration of Independence and, subsequently, the Constitution, which expressed the sentiments distilled from the institutions growing out of the independence of the pioneer.

The Declaration expressed a generous belief in political and social equality which, while not characteristic of all institutions of the period, represented a general consensus and an insight into the ideals for the future. The Constitution and the Bill of Rights expressed the astounding ideas of restraint upon government and the supremacy of the individual citizen. While, of course, nothing in the Constitution authorizes a pluralistic or free enterprise system, the idea that all powers not delegated reside in the people implies a freedom to engage in enterprise and do it in a diversified and individualized way.

Thus, under the aegis of a new political arrangement expressing the common sentiments of the time, the experience which had been called "colonial"—obviously not semantically an appropriate word to characterize post-Revolutionary days—continued to unfold. But to consider the new era-to-be disjointed from the past would be even more incongruous than to continue to use the word "colonial." The political upheaval notwithstanding, the same educational, scientific, commercial, religious, and social institutions continued, and they continued to proliferate and modify and change as more people were involved and more individual differences had to be accommodated.

In this continuum of institutions, however, the age of invention erupted almost simultaneously with the founding of the new nation, and the new nation had to deal with the issues created by scientific innovation unprecedented in human experience. The advent of mechanization characterized all Western nations, but in the United States Eli Whitney's cotton gin was followed by Fulton's steamboat, and then by the Erie Canal and the development of the railroads. A nation of individual farmers, artisans, and tradesmen was faced with the need to organize vast numbers of people to accomplish the huge projects which flowed from the development of engineering science. The social form which emerged to provide the means of amassing capital and the labor required for gigantic enterprises was the private capital corporation.

Whatever the reasons for success—the tremendous natural resources, the energy of a new people, the response to democratic life, the continuing growth of invention—the 175 years of post-Revolutionary experience brought on a golden age of marvelous prosperity and constantly expanding human experiences. The Industrial Revolution and the urbanization of society changed the whole aspect of the nation; yet through all eras of change, the concepts of the Revolution and the institutions which were born in colonial times and translated into the new nation remained constant points of reference. We continued to take pride in ourselves as a nation totally committed to the justice of democracy.

The relative importance of educational, scientific, commercial, religious, and social influence changed from time to time. But behind all the change was the abiding sense that people developed their lives in accordance with their own insights and predilections. Governmental order did not mold society, and governmental institutions were not unlimited or self-perpetuating.

On the other hand, such tumultuous development could not take place without conflict and abuses. Out of the imbalances in power, which from time to time created hardships and glaring inequities, came the recognition that where abuses were defined, the public interest was superior to the private. Excesses and abuses of private powers were curbed and checked by governmental intervention.

In the closing decades of the 19th century, both the state and federal legislatures and the Supreme Court began to consider the novel social and legal questions produced by an emerging industrialism. These questions were increasingly concerned with the economic impact of groups and corporate entities on society.

The problem arising out of this emerging industrial society produced two governmental responses. One was to create additional criminal and civil penalties to prevent undesirable conduct. The antitrust laws of the 1890's illustrate this type of response. The other response was regulation. And on both the state and federal levels, a number of regulatory commissions were formed to regulate the railroads and the emerging utility industries.

At first, such regulation was simple, understandable, and uncomplicated. Today, governmental regulation has grown to monstrous dimensions, and the complexities of modern life have created totally new issues and challenges. We continue to say that we have ideals of individual freedom, but whether our governmental institutions are really protecting these human rights may be debatable. The question is no longer Lincoln's query whether a nation dedicated to democratic ideals can survive, but rather a question of whether individual human beings can remain free in a society which has become increasingly interdependent and dominated by social regulation.

We're so interdependent upon each other now that it is difficult to imagine a person living alone and sustaining himself. No longer can we make our own shoes, build a cabin, hunt for our own food, or do all those things which a person 200 years ago could do for himself. We have to depend on others for our food, our transportation, and all the other necessities of life. But in the interdependence which we have acquired there are great dangers.

It is the *generalized* thesis of this book that human beings can remain free only if their institutions, and the organizations and associations in which they live their group lives, can remain free and are not dominated by the government.

It is the *specific* thesis of this book that the public utility company is a unique crossroad of public and private institutions suggestive of a way in which private creativity can flourish amid a constant concern for, and in response to, the public interest.

To support this thesis, it is necessary to examine the utility, not stereotypically, but as a function of, and indeed a product of, its community. A utility company is a social mechanism which has evolved out of history and is a response of a democratic free society to the problem of supplying an essential service for modern life. While its critics too often condemn it as a predatory organization seeking unjustified profit, abusing its monopoly position, spoiling the environment, using up natural resources, and charging too much for its services, their complaints are really irrelevant as directed against the utility company itself, because that company is only balancing out the various demands and impacts which are made upon it. If the criticisms are justified, they are criticisms which should be directed against society as a whole and the values which society seeks to emphasize and cultivate.

The criticisms should be weighed against the great benefits, not only technological but also sociological, which the utility mechanism provides. The utility company has the form of a private business corporation, and provides all the great benefits which that social device bestows. But the utility company also takes account of the public interest through the vast interplay of the social forces which affect it, both governmental (represented by the intense regulation to which the utility is subject and, increasingly, legislative pressures) and private (represented by consumer, environmental, and other groups). By the unique combination of a private and public character, the public utility corporation provides to the public the attributes of an independent entity grounded in free enterprise but limited by public regulation. With its roots in a private property enterprise, free from the political organs of society, it is not subject to direct political pressure. And with its independent organizational structure, it is not involved in the bureaucratic mechanism of government. Yet at the same time, it is so subject to governmental reviews and public disclosures that it is truly a public operation.

The following pages explore these issues:
(1) The attributes of the private corporation—for without this consideration, the utility cannot be understood.
(2) The interactions of the private utility with the groups of people related to it, which in effect describe the utility operation.
(3) The technological and social benefits that a utility operation generates.
(4) Some attributes of public agencies contrasted to those of the utility company.
(5) The basic problem of regulatory methodology.
(6) The interaction pattern of the utility as a prototype for developing the free institutions, and the associated schemes for economic development in the future.

Out of this pattern is seen a product of the American genius, that wonderful ability of a diverse people to put things together out of their own initiative and creativity. The result is a supply of utility services, unique in the world and in history, with manifold and hitherto unimaginable benefits, and the development of a large, successful, dynamic industry operating through many relatively small companies responding to and supportive of local interests.

The story is remarkable, and its future may be even more satisfying and rewarding.

Chapter 3

HOW TO ORGANIZE THE NONGOVERNMENTAL GROUP LIFE

Before one can talk specifically about the utility itself or the utility process, it is necessary to consider some characteristics of the form of organization the private sector utility uses to carry out its business.

Private sector utilities are usually organized as corporations, as indeed are most large-scale and even small-scale enterprises. In the dialogue in newspapers, magazines, radio, and television, the corporation is often condemned and criticized. It is seen as a greedy and negative force, dominating our lives and depriving us of freedom of action. The truth, however, is that the corporation is the form which society has devised to provide the way of distributing goods and services in a non-monolithic society.

In a governmental and social structure which permits free independent activity, economic enterprise will be carried out by individuals as long as the society remains relatively simple. When the society becomes industrialized and interdependent, however, most economic activity can be carried out only in large-scale enterprises which have the capacity to amass large amounts of capital. Under these circumstances, economic activity has to be carried on by some kind of group formation, and the form in our society for such activity has been the economic corporation.

We can imagine that in a primitive society, people sat back and asked: "How are we going to organize?" One way was to set up the organization with a king or chieftain at the head who told everybody what to do. Or the other way was to let the people do it their own way.

This "Let them do it their own way" approach is precisely the way in which the American process unfolded. As economic life became more complex, group enterprises evolved from simple proprietorships, to partnerships, to corporations.

But when you follow the "letting them do it their own way" approach and form these groups, you have to define the rights, powers, and the disabilities of these groups. This defining of rights, powers, and disabilities is very complicated, but the same defining has gone on for centuries in the English common law for individual persons. Most people don't stop to think about the rights of persons to own and sell property, for example. They take it for granted. But all of these individual rights have been developed through a long process of evolution arising out of separate experiences and separate cases, and the details of individual rights can become very complex indeed.

But people do not exist very much in the individual context. In the famous words of Martin Buber, "All real living is meeting."[1] It is in this meeting that human life unfolds, and so human beings seek and constantly form groups.

The complexities of modern life require group living. We have long since stopped being independent individuals who can feed, clothe, and shelter ourselves. Diversity of talent and occupation, and the need for specialization, require that we live closely together and cooperate and interact with each other. Group living is an attribute of modern life.

But how are groups of people formed who have common interests and common goals? How does society regulate their conduct, both internally and externally? How does it provide for their continuance or their demise? How does society apportion equity among those who comprise the group? How do individuals make decisions? How do they carry them out? How does society make the group responsive to the largest social requirements?

The corporation deals with these problems, but the process is far from simple. The literature relating to the corporations is complex and often esoteric, and the issues in which it is involved require a sophisticated degree of scholarship. At the same time, the complexities of corporate law, corporate finance, and corporate accounting tend to obscure some of the basic principles.

In considering the process of group life, or the process of the corporation, two issues emerge. The first issue is: How do you deal with the group as a group? What are the relationships of the whole with other groups, with other people, and with the community? The second issue is: How do you deal with relationships among the members of the group?

The first issue deals with the corporation as an entity. It deals with law, economics, finance, and accounting. The second issue deals with people. Whatever abstract views or sophisticated arguments are developed, the fact is that the corporation is a social form to organize the activity of people—people who have varying degrees of attachment and contact with the corporation and who perform diversified functions. It is a form of associated activity bringing together the three common interest groups of people; namely shareholders (or more broadly to include holders of debt and investors), employees (including management), and customers. To the corporate officer, the corporation may be the total preoccupation of his life; to a shareholder, it may represent a casual investment, of no interest beyond the market quotation of his stock or dividend rate. To the customer, it may represent an infrequent contact as now and then a product is purchased. Yet all of these people are involved in a web of relationships which prescribes modes of conduct, attitudes, and value considerations.

The next two sections deal first with the corporation as an entity and second with the corporation as an aggregate of people. The dichotomy is often perplexing and, in a sense, contradictory. And yet, when corporate issues must be considered, they always must be viewed with this bifurcated understanding in mind.

I. The Corporation as an Entity

No matter how much a desire for humanism in business relationships would lead us to focus our attention on the nonlegal and non-economic elements of the corporation, it cannot be avoided that the form of the corporation is that of a separate, discrete legal unit or entity. And while this idea of an independent social entity does not seem particularly unnatural or unusual, in truth it represents a sophisticated thought process which was only developed after an extended evolution.

For a long time, groups were thought of only as an aggregate of individuals without any inherent characteristics of an association. It seemed to be a difficult concept to grasp that the group *qua* group could have legal and social relationships and interactions. A brief excursion into some ancient legal history gives an insight, not only of the problems themselves, but also of the typical response.

Interestingly, as far back as 1292—200 years before Columbus's main exploit—an early dispute is recorded concerning the meaning of a corporation. Interestingly and perhaps prophetically, it arose out of a problem of utility rates. While 1292 is remarkably remote, even then utility issues were at the "cutting edge."

In an old case, an abbot in charge of an English church objected to an enterprising sheriff's decision to impose a toll on the monks for using a road. The abbot said, "You can't charge a toll because I'm a religious corporation," and he brought a suit on behalf of his group of monks—his corporation, as it were—but was told, however, that this procedure could not be followed. The court said: You're just a group of individuals—Monk A, Monk B, Monk C, and so forth. If there were a complaint, it would be by each monk as an individual but not as a group.

This must have been very frustrating to the abbot, who was trying to say that the church was immune from such mundane charges. But the court said there is no such thing as a church in that sense. It is just a group of monks, and they have to pay tolls just like anybody else.

Nevertheless, the conclusion of the case was in accord with the views not only of the time, but also of a long time to come. In 1292, when this case arose, only a few imaginative souls thought a group had legal rights. But people continued to form groups and develop associated activity in which they joined talents and resources. As this process went on, disputes and controversies developed.

Throughout this early period in the history of English institutions (from which ours derived), in which the modes of dealing with group activity were so uncertain, land was the only important form of wealth. The most significant area of the law and, indeed, of public affairs was that dealing with property. The significance of the law dealing with contract rights and tort claims was very limited. As a matter of fact, personal property, which today is the most important type, especially in the form of corporate securities, was of only minor importance, consisting for the most part of tangible things—furniture, jewelry, etc. The law of property, rivaled only by the criminal law, was the predominant consideration of the courts.

It is basic to a system of private property, as opposed to communal property, that property be held by rational and responsible units. Private property implies that the holder of the rights will have certain capacities which will enable him to make proper use of the property entrusted to his care. The owner of property must be one who can make decisions on the disposition of the rights he holds.

Hence, a property holder cannot be a minor, a lunatic, an imbecile, an incompetent, or anyone who lacks the decisionmaking ability that would enable him to employ the rights which he holds. In fact, surprisingly, for a long time, married women were not permitted to be owners, for their wills were considered to be those of their husbands, and it was felt they lacked the independence of decision, a characteristic considered essential to make them suitable property holders. This inequity has been corrected—as cogent an illustration of social progress as any.

Since the holder of property rights will be involved in disputes concerning these rights, he must have the capacity to sue and be sued. This involves the ability to prosecute claims and to defend against demands. It also implies the ability to satisfy a judgment of a court.

The capacities the holder of property must have in order to function within the philosophy of the legal system of the law of property are ideally provided by the individual. He can make decisions, he can defend himself in the event of disputes, and he can be held accountable for his acts. When questions about property matters arose it was very logical to recognize how easily these questions could be answered if the corporate group was considered as an individual.

In our American-English legal system, an individual has the right to decide. He doesn't have to ask anybody what his decision should be; he can make it himself.

English law considers the individual to be the basic social unit, and while this may seem natural to our way of thinking, it is not the only possible legal consideration, nor a necessary one. Early Roman law, for example, looked to the family as the basic jural unit. Roman law thought of an individual as part of the family. But English law is not like that. The individual himself is the prime thing.

In this system of individualism, it appears almost inevitable that individualism would be applied to groups and that groups of people would be treated as a unit. While this approach seems so obvious to us today, the development of the idea did not come quickly. Certainly there were other possibilities and hints of an alternative method, suggested by

Roman law, in which the corporation would be thought of as an "artificial family." This appears in some early cases.

But even the semantics of the word "corporation" has a meaning. Derived from the Latin word "corpus," meaning body, the idea emerged that the corporation was a "body," a "somebody," if you wish.

And so, a brilliant leap in conceptual thinking was made, and the courts expressed the view that the corporation is not to be thought of only as a "body" but as an "artificial person." By the invention of the artificial person, the courts were able to apply all the law, tediously developed through centuries, to the group. In one brilliant conceptual leap, a tool and a mechanism for handling groups were suddenly unfolded. In one jump, the entire body of law pertaining to a person became applicable to the group.

This is a logical explanation of the mental process and it turned out to be the way the concept developed historically. A number of cases deal with the issue, but two of them—*Sutton's Hospital* case, rendered in 1613 by the famous English jurist Lord Coke, and the *Dartmouth College* case (officially *Dartmouth College* v. *Woodward*) decided by the United States Supreme Court in 1819—are usually cited by legal scholars as definitive statements of the character of a corporation.

The language in both of these cases is interesting for the quaintness of expression and for the insight which the quaintness itself supplies.

In *Sutton's Hospital* case, Lord Coke said: ". . . The corporation itself is only *in abstracto,* and rests only in intendment and consideration of the law; for a corporation aggregate of many is invisible, immortal and rests only intendment and consideration of the law."[2]

And in the *Dartmouth College* case, Chief Justice Marshall said: "A corporation is an artificial being, invisible, intangible, and existing only in contemplation of law."[3] In the same case, Justice Story said:

> An aggregate corporation at common law is a collection of individuals united into one collective body, under a special name, and possessing certain immunities, privileges and capacities in its collective character, which do not belong to the natural persons composing it. Among other things it possesses the capacity of perpetual succession, and of acting by the collected vote or will of its component members, and of suing and being sued in all things touching its corporate rights and duties. It is, in short, an artificial being, existing only in contemplation of law and endowed with certain powers and franchises which, though they may be exercised through the medium of its natural members, are yet *considered as subsisting in the corporation itself as distinctly as if it were a real personage.*[4] [Emphasis supplied.]

This assumption by the corporation of the capacities of the individual has two important consequences:

I. The first relates to the legal procedures which enable process of law to be applied to the corporate group. The group can do things, undertake activities, and enter into relationships, all of which will be enforced by the courts. And this enforcement, or potential of enforcement, is the ingredient which makes the whole system work.

In the corporate world, the things which the group—i.e., the corporation—can do are expressed as its powers. These powers of a corporation, as set forth in corporate statutes, may seem to be a technical matter of concern only to lawyers and other corporate technicians. One thinks just because the powers are written in a statute, they are very technical. But they are very illuminating also. In truth, the powers a corporation possesses describe its nature as an entity by defining the things it can do. In short, studying the powers a corporation possesses is a very handy way of defining what it can do.

Corporate powers demonstrate the capacity of the corporation to operate freely in a legal system whose basic tenet is that the natural person is the source from which all powers come. The government is only the secondary source. Thus, for example, when the powers of a New York corporation, as provided in the Business Corporations Law, are examined, the functioning of a corporation as an entity is precisely the functioning of an individual.

Summarized, the law provides that each corporation shall have the power to:

(1) Have duration.
(2) Sue and be sued (as natural persons).
(3) Purchase, own, hold, improve, employ, use, mortgage, and otherwise deal in property.
(4) Make contracts, give guarantees, incur liabilities, and borrow money.
(5) Lend money.
(6) Do business.
(7) Appoint agents.
(8) Make donations for the public welfare or to the community fund; hospitals; charitable, educational, scientific, or civic groups; or similar organizations.
(9) Pay pensions and carry out pension, profit-sharing, share bonus, share purchase, share option, savings, thrift, and other retirement, incentive, and benefit plans.
(10) Be a promoter, partner, member, associate, or manager of other business enterprises or ventures of other corporations.

II. The second consequence of the corporate group—assuming the capacities of the individual—relates to the basic premise of our society. The corporation is a creature of law, established by the State. But once established, it possesses legal and constitutional rights, and they are those legal and constitutional rights of the individual who stands supreme in our fundamental way of looking at things.

The corporation, to the extent it is likened to an individual, stands independent and unassailable in specified particulars from the power of the State. Its *persona* quality is basic to the constitutional thinking about a corporation. Even though the precise language may not be used as frequently as it was in the past, the nature of the corporation as an artificial person was the foundation of decisions by the Supreme Court which afforded the corporation some of the privileges, rights, and immunities set forth in the Constitution.

Thus, a corporation was considered to be a person within the meaning of the Fifth and Fourteenth Amendments of the Constitution. For utility companies, this protection is probably the only restraint against confiscatory regulation.

In the late 19th and early 20th centuries, the constitutional posture with respect to corporations may have been too protective, and the role of the legislature was frustrated by constitutional barriers that were not in phase with economic and social reality. This issue was debated over and over again in the cases which passed on the constitutionality of a number of the legislative enactments of the New Deal during the 1930s.

Under these cases, the freedom of the corporation was sharply curtailed, and the right of control by the federal government vastly expanded. There is little substantial debate today concerning the validity of these New Deal decisions, but the climate of the 1930s in which these cases were decided was vastly different from that of the 1980s.

The New Deal limitation of corporate constitutional privileges occurred when the impact of governmental agencies was trivial. But today, the vast and continuing proliferation of government agencies, and the extension of those already in existence, creates a totally different society. In view of this background, it is time to take a new look. Indeed, the issue directly becomes the one stated in the previous chapter on the utility framework: whether individual human beings can remain free in a society which has become almost morbidly interdependent and perilously close to being dominated by social regulation and thought control.

In this context, the entity character of the corporation needs to be reestablished and the old concepts reaffirmed, not because they have any virture arising out of their antiquity, but because the rationalization of the corporation becomes a key issue in the incidence of a free society. The corporate entity concept is justified in terms of how it actually functions, the fundamental and dispositive social test alluded to in the preface.

Before the New Deal, there was a very tight limitation on what the legislature could do to impose corporate restrictions, because such restrictions could result in the deprivation of property without due process. After the New Deal, a lot of things were accepted and the ancient thoughts of a corporation's constitutional rights were easily disregarded by rationalization concerning social policy consideration. But the basic concepts covering the corporation, namely how groups are organized and how they function freely and fairly in a pluralistic, non-monolithic society, need to be reasserted. The modern problem is acute and needs to be dealt with in terms of basic corporate considerations, which tell us that the corporation provides an independent decision-making, activity-producing, power-possessing entity which is a part of the community, but not part of the state or of the government.

The entire lesson of the modern corporation is that it provides a sound, secure way of organizing human activity while safeguarding protection of human and individual rights which have been forged out of the long process of history. Walter Lippman once stated that:

> On American democracy: It has worked, I am convinced, for two reasons: The first is that government in America has not, hitherto, been permitted to attempt to do too many things; its problems have been kept within the capacity of ordinary men. The second . . . is that outside of the Government and outside the party system, there have existed independent institutions, and independent men . . . the judiciary . . . the free churches, the free press, the free universities, and no less important to the preservation of democracy, free men with sufficient secured property of their own, farms, factories, shops . . . protected by law and not dependent upon the will of elected or appointed officials . . .
>
> On "the deepest issues": The deepest issue of our time is whether the civilized people can maintain and develop a free society or whether they are to fall back into the ancient order of things, when the whole of men's existence, their consciences, their sciences, their arts, their labor, and their integrity as individuals were at the disposition of the State . . .[5]

If to that list of free groups we add the free corporation, then his quote expresses in eloquent terms the theme being discussed.

II. The People of the Corporation

Notwithstanding the legal nature of the corporation and the mumbo jumbo that goes with it, and without detracting from its legal nature, the simple fact which must constantly be before us is that the corporation is comprised of people. When it operates in day-to-day life, the corporation is, in effect, a framework of institutions which provides for human activity within a complex web of relationships among the people.

But the institutions and the relationships do not exist until they are activated by people, and it is these interactions of the people which produce the reality of the day-to-day operation of the corporation. In this view, the corporation is the institutional process which (1) coordinates the resources of people and capital to supply a product or a service and (2) activates the stream of economic flow which involves the sellers-buyers market and the return on investments.

The people of the corporation are shareholders or investors, management (including the board of directors and officers), employees, and customers. The interplay of the groups within the corporation, the safeguards which have been created to protect the interests of individuals within these groups, the modes of the organization of the group, their methods of functioning and other characteristics, all present a fascinating panorama. But the purpose of this study is by no means to deal with the corporation generally or to analyze the characteristics of shareholders, management, employees, and customers as they function in the general corporate structure.

It is, however, the purpose of this study to see people in the public utility perspective, and also to see the private sector public utility as a functioning entity contributing to the processes of a free society. And so with these preliminary comments about the corporation, we turn to the utility corporation itself. Utilities, among many other things, tell us in a most graphic way how these groups of people—i.e., investors, management, employees, and customers—interact and supply benefits to each other. The next chapter deals with the interactions among the groups which comprise the corporation, and describes the utility company in these terms.

Chapter 4

THE PUBLIC UTILITY

John Donne's famous words that "No man is an island, entire of itself . . ."[1] refer to our interdependence upon each other as human beings. Properly paraphrased this quote would be an accurate reference to a utility company.

A utility company, in a teeming world of interacting elements, is a specialized corporation. It has all the attributes of the corporation as described in Chapter 3 and is one of the elements in a free society which, having a structural existence of its own, can operate as a discrete entity within society.

A utility, however, differs from the usual business corporation. Besides the interaction within the corporation, it engages in a vast and continuous interplay with the community. It represents the public interest—not by itself, but as a result of the complete process in which it engages. It is a product of the evolution of a social form. It is in a sense a progeny of the corporation. It is the latter-day embodiment of the corporate idea interfacing with other groups within a governmental context which is no longer laissez-faire.

Far from being an outrageous relic of the past, stifling progress by representing vested interests, it has within it the potential to be the prototype of future social forms. This point will be expanded in Chapter 10.

A utility company is a complex organization, not only within itself, but, more significantly, in terms of the many external relationships with which it is involved. It cannot be viewed alone or in isolation but must be considered a part of the complex web of interactions of social forces.

First, like any other business, a utility company has investors, customers, and employees. Serving the interests of these groups is its main area of activity and the reason for its existence. Second, a utility company is subject to intensive regulation by governmental agencies. Third, a utility operates as part of a vast interconnected utility system. In fulfilling this function, it participates in many cooperative industry efforts which ensure the reliability of electric power, such as power pools and regional councils. In fact, since utilities are noncompetitive with each other, they cooperate and interact with each other in a way no other business does. Last, a utility company is subject to the economic, political, and social conditions which characterize our national life and which have so much to do with the degree of its success.

A great deal of literature on the economic, engineering, regulatory, legal, and ratemaking aspects of the utility company has been produced, but very little or none has been written about the utility as a social organization or the implications of its interactions with the community it serves and society generally.

If the utility is viewed in the context of these various functional dimensions, a societal view of the utility emerges, which can lay the foundation for an understanding and appreciation of the utility as it functions in society; a societal view can provide the basis for rational legislation affecting energy needs.

The Utility Company as an Entity

Notwithstanding the need to look at the utility company as a composite of social groups interacting in a continuous flow of transactions, it should be recognized and reaffirmed that a utility is a unique and separate entity. It is a business founded by entrepreneurs who, representing investors, saw an opportunity to employ resources and receive an attractive reward. In pursuing their own objectives, subject to all the limitations of economic and legal restraints, these entrepreneurs initiated the enterprise which produced many social values.

A utility company is a corporation formed under the laws of the state in which it operates, with all the attributes described in the previous chapter. It is, thus, an artificial person with the legal and constitutional rights, privileges, immunities, and obligations of a natural person.

The utility company, then, is an entity and, once created by the sanction and through the operation of state law, it has the rights prescribed in corporate law to hold property, to make contracts, to sue and be sued, to be liable for debts and taxes, to continue to exist despite changes in its personnel and the membership of the groups which are involved with it, and to act totally as if it were an individual.

As an entity, the utility company has the right and responsibility to act in its own capacity apart from the groups which comprise it. Its internal mechanism for making decisions is prescribed by statutory law. The utility company has many of the legal and the constitutional rights and immunities of a person subject to the law.

In light of its corporate and entity characteristics, the utility company has the right to make decisions and take actions in accordance with these decisions, to express opinions, to advertise, to take positions on legislative and judicial matters, and, in short, to have that independence of a separate social body which is one of the cornerstones of a free society.

The Utility Company as an Organization of People

When it operates in day-to-day life, the corporation is, in effect, a framework of institutions that provides for an interchange of human activity within a complex web of relationships—even though it is a legal entity. For example, in a corporation, the chairman of the board interacts with the president, and he may interact with the treasurer or financial vice president, who then works with somebody else. No matter what one does, it always has some kind of reaction.

A utility company is, in many ways, a sum greater than its parts precisely because it is composed of people who perform specific functions.

And the reality of the day-to-day operation of the company is the interactions among the people both within and outside the corporate structure who in one way or another have dealings with the utility. The utility corporation is an institutional mechanism which coordinates the resources of people and capital in order to supply utility service in a designated area.

The utility company was formed by investors and entrepreneurs to obtain a return on capital. Once formed and endowed with franchises, it exists to satisfy the interests of investors and customers. It strives to supply utility service at the lowest possible cost. Yet at the same time, it tries to provide a reasonable return to investors—certainly one which will be sufficient to attract new capital.

The employees and officers of the utility are engaged to carry out its services to the customers and investors, but as time goes on they become a group whose interests are served by the utility itself. While theoretically it is true that the utility company is formed for the investors and customers, in reality it exists to serve the interests of the following groups: investors, customers and employees, and the community as a whole. Each of these groups acts in accordance with its self-interest, and interacts with each other in a constant flow of transactions and relationships.

Investors

Historically, the enterprise is started by the investor, and to the extent that the customer group demands additional service, the enterprise is "restarted" by an influx of additional blocks of capital. The investor's role is to supply the capital funds for the enterprise and to insert money into the stream of technological and human activity which the utility represents. The investor invests his money in the utility so it can build the plant which serves the customers, with the expectation and the constitutional entitlement of a return on the capital funds which he has invested. But, unlike the customer, he is not captive. He can pick and choose in the open market into which particular enterprise stream he wishes to place his money.

He must be attracted to the enterprise and to its potential. The basic test of the amount of money which the captive customer must pay is the amount which, after taking into account all the requirements of the enterprise, will attract the investor. But without that attraction, the investor can withhold his funds or demand higher than usual returns.

In the utility business, utility equity is issued based on the following assumption: the utility shareholder, because of the security of his investment, accepts a low growth but a high dividend payout (i.e., a high yield and minimum appreciation of his stock). Still, the security of his investment depends on the regulatory climate which recognizes that the customer pays the entire cost of the business, including the cost necessary to attract present and future capital.

This understanding has worked very well in the past. However, as fuel costs have risen, regulators have tended to restrict the cost which the customer has to assume. If this process continues, the shareholder-utility relationship will totally change, and the shareholder, before he supplies equity, will want assurance that he will obtain the same benefits which the investor in a competitive business expects—namely, high return and high appreciation.

Aside from the obvious roles which investors play in supplying the funds for the enterprise—all of which have been and are continually being analyzed in economic studies—the legal rights of investors and the ability to undertake certain acts to which their ownership entitles them creates very subtle pressures and incentives which deeply affect the conduct of the enterprise and contributes to its stability.

None of these pressures were contemplated in the original design, or the statutory design, of the corporation; they evolved from the sociological and interactive role of the corporation. The process of forming a corporation is a wonderful example of how social forms unfold and develop, regardless of the master plan or designs, as people contrive to meet practical situations.

Exertion of such pressures through the shareholders arises as follows: A shareholder is a person who owns a shares of stock in a corporation. Such ownership gives him certain rights which may be divided into two classes: (1) the right to vote in the election of directors and on all fundamental policy matters, and (2) the right to receive dividends and to participate in distribution of the assets of the corporation, should it be liquidated.

These two classes of rights pertain to two different functions. For the purposes of the first class, the stockholders are an aggregate of people with a decisionmaking function; for the second, they exercise a proprietary function whereby the stockholders act as individuals who respond to economic attractions and incentives.

In exercising the first class of rights, the stockholders, as decisionmakers, theoretically are required to form a face-to-face group when voting. But in the modern corporation, the shareholders' meeting does not serve as a valid process of decisionmaking. The need for using proxies changed the nature of the corporate meeting. No longer does one expect the collegiality which was expressed in the early 1800s by New York's Chancellor Kent, an early and imposing formulator of corporate law, in describing a corporate election:

> The election must be given when the members of the corporation are duly assembled collegialiter; and they must act *simul et simel* (simultaneously) and not scatteringly, and at several times and places.[2]

In fact, quite the contrary is true; the shareholders do act scatteringly.

Yet, notwithstanding the fact that the proxy system destroyed the decisionmaking function of the corporate shareholders' meeting, the shareholders, through the meeting and otherwise, exert very powerful pressures.

The first pressure arises out of the annual meeting, which has become an institution leading to publicity and to political action. While anti-management proposals in proxy statements by and large have not been adopted by corporations, the raising of these issues and the concomitant stimulation of public interest and opinion have forced management to give the proposals serious consideration.

Current Security and Exchange Commission proxy rules make the ability to raise issues in this fashion an even more powerful influence on the corporate conduct. For example, a utility shareholder places an anti-nuclear proposal on the ballot, a shareholder of an oil company places a proposal concerning company investment in Chile, and a shareholder of a chemical company places a proposal for scientific investigation of the effect of the company's products on the environment. All of them know that the proposal hasn't the slightest chance of passing, but the issue is raised in a public forum, the press gives it coverage, and a pressure is imposed upon management which no amount of letterwriting or personal meetings could ever accomplish. As a result, no management appears before a shareholder group without a rationalization for its position, and the audit mechanism which the process of the meeting provides has an influence over corporate practices pervading throughout the entire corporate year.

The second pressure which shareholders can bring is the threat of a shareholders' suit. While only a few instances actually reach litigation, management's thinking, guided by legal counsel, must always take into account the potential of shareholder action. Management discussions concerning corporate policy always consider the test of what a shareholder's suit could do to any proposed course of action. It is like a Banquo's ghost at every corporate conference, the threat itself is quite sufficient to provide a protection of the shareholders' interest.

The prime role of the stockholder, however, is to provide the equity component of capital. In the property-owning role, the shareholders are the representatives of the financial concept of equity. In a utility which is constantly in the financial markets, this takes on a special meaning. In addition, in terms of functional considerations and the importance of cash flow in an enterprise, equity protects the bondholders; the amount of equity is the margin of safety which an enterprise has available before it must be in default. In short, equity is the measure of distance between the enterprise and the bankruptcy court.

The growing use of governmental agencies and authorities operating enterprises illustrates the importance of this. In an organization which has no equity component, all revenues after expenses are committed to the payment of bond interest. Accordingly, a means is required whereby rates to consumers can be immediately increased as expenses are increased. There must be an automatic passthrough of costs; customers of public authorities are not afforded the luxury of a rate case. Thus, toll road, bridge, tunnel, railroad, utility, and other authorities increase their charges in sham proceedings in which the public either has no voice or, at best, only a participation.

And, if notwithstanding the automatic passthrough, if revenues fall off due to decreasing volume of business, generally unfavorable economic conditions, or otherwise, the payment of bond interest is immediately jeopardized. In an enterprise with an equity component, when revenues do fall off the dividend may be threatened, reduced, or eliminated, but the interest payments are not immediately threatened. Indeed, the enterprise has a time period or reprieve during which it can recover revenues and take remedial steps to restore its capacity to pay. The shareholders thus provide the enterprise with a great deal of stability.

If Consolidated Edison Company, during the period when its revenues declined because it did not pay a dividend, had been a governmental enterprise without shareholders and a preestablished recourse to general tax revenues, the decline would have affected interest payments. Such an enterprise may have been in default and, possibly, in bankruptcy. Thus, the higher return which customers provide through their payments for the benefit of shareholders is not extracted, in the colloquial view of things, for the benefit of shareholders, but is an appropriate compensation for providing the equity source which gives financial protection and stability to the enterprise.

Bondholders, too, in seeking to protect themselves through covenants in trust indentures, are acting as a class to increase the stability of the enterprise. A coverage requirement, for example, provides the margin of safety (something like the existence of the equity fund) against declining revenues, and serves as a pressure upon management and regulators to insist that revenues remain sufficiently high in order to satisfy the interest requirements of the bonds.

If, for example, the West Side Highway in New York City, which collapsed because of improper maintenance, had been run by a private toll road company (which, for other reasons, it could not be), maintenance funds could not have been diverted to other purposes, but would have been required by the trust indenture to be applied to maintenance work. If the funds had not been so applied, the bondholder group could have required it through litigation.

Finally, the investor group should be recognized as a composite of other groups generally regarded as institutional investors—pension funds, trust funds, mutual funds, insurance companies, banks, etc.—all of which serve as a conduit between the company and the vast numbers of people these institutional investors represent.

Through these processes, and many others which cannot be described in a short chapter, a complex web unfolds reaching to countless people who become a part of the entire corporate scene. And by the evolution of the corporate structure and the dynamic quality of capitalism, investors give the corporation, and more specifically the utility, a broad democratic base.

The Customer Group

The relationship between the company and the customer is an economic one since it involves the exchange of money for a service. The customer seeks the services of the utility company and finds them necessary and desirable—and for that he is willing to pay money. Once the organization is established by investor money, the customer initiates the flow of that money and starts the chain of reactions. The customer is not, however, able to make a free choice because, to a degree, he is a captive customer; he must deal with a specific supplier—his local utility—and his choice is limited to determining to what degree he will demand service. He can live simply with small demands or he can use all of the appliances and devices he wants, in which case he will use increasing amounts of utility service. He also can use alternate fuels or energy services, principally for space heating or industrial processes.

Because the utility customer is geographically bound to the company which serves him, his relationship to the utility is not unlike his relationship to his local government. With local government, he has no choice but to accept its services and to pay taxes. Likewise, he must accept utility services and pay its rates. Viewed in this way, the utility is a monopoly—its customers have to deal with it—but also the utility is democratic because it must deal with everyone in its service area. It cannot pick and chose its customers; it must be available to the whole community.

This quality of the relationship between the utility and its customers is political, in the generic sense of the word, and the utility must seek to deal with it with a political consciousness. The customers of a utility are like voters in a political district. Just as the politician has to seek the favor of the voters, the utility has to deal with its customers. Thus the relationship might be said to be political because the utility depends on the goodwill of customers.

This does not mean that the utility engages in political action in the typical sense. The utility must not engage in politics, but it must use the techniques of politics. In short, it must be responsive to its customers and it must seek their acceptance and understanding. Above all, it must find grounds for identifying with its customers.

Likewise, from the customer's side, his recourse is political rather than economic. He may move through the regulatory process, participating in proceedings of the regulatory agency. In his role as a citizen, he can exert political and public pressure. He can express his views in newspapers and on television and radio, and seek to have his way through the concerted effect of public opinion.

There is no escape for him, however, from paying his share of the full costs of electric service, and the cost of this service is going to increase as long as prices and capital costs increase. Although there is a great deal of research on new and, in some cases, exotic ways of supplying energy, there are no apparent substantial improvements in the state of the art, or in any techniques which will reduce increased costs of their development. And if conservation, mild weather, supply shortages, or other reasons cause the growth in sales to be less than the growth in costs, unit costs are bound to increase.

Furthermore, in addition to the full technological costs, customers also will have to bear, as they have in the past, a large component of social costs which primarily is the payment of taxes. The extent to which utility companies contribute to the general welfare by paying taxes is usually not recognized, yet this exists and will continue to exist. In addition, the customer will have to bear heavier environmental costs. Utilities have been criticized for not taking sufficient measures to preserve the environment. Yet, they have been selling electricity or gas at prices which do not reflect the very substantial increasing costs which are required under environmental regulations, and their rates have been inadequate to cover these costs. Future costs will continue to consist of technological, social, and environmental components, with the last two probably increasing.

Customers also must pay the social cost of contributing to capital formation because no economic transactions should be allowed to escape the need to contribute to society's pool of savings.

The practical lessons of this discussion are that the utility must recognize more emphatically the role its customers play and the political and economic consequences of its relationship with them. Adversary approaches must be minimized, and an effort must be made to take rate cases out of the adversary forum. Even the nomenclature of a "rate case" is anathema and suggests a trial. Instead, it should be an inquiry into facts and an exercise of an audit function in which the public and the regulators are involved.

On his part, the customer must recognize his responsibility and not irresponsibly engage in name-calling or accusations of bad faith when the increasing costs of society are reflected in electric rates. He is entitled to good service at fair rates, but not a free ride.

Employees

Employees expend their time and talents in operating the company in return for wages. While the enterprise was not established for them, they make a greater investment in it than anyone by devoting to it the major part of their productive lives. In a real sense, they are the enterprise, and they mold its character and its performance.

The employees bring to the utility the expertise and skills necessary to operate a technical business. It is their skills in operating extraordinarily complex machinery which converts the raw resources into a usable energy form and provides for its distribution. The skills range from professional and highly trained capabilities to reasonably routine ones, which nonetheless require education and training. All of this is mostly taken for granted, but it is truly the end product of a complex process. Sometimes this coordination of education and work activity is difficult, and brings up all of the problems of interdisciplinary communication.

Part of the problem in our modern world is that doctors don't understand lawyers, lawyers don't understand engineers, engineers don't understand accountants, and accountants don't understand salesmen. And there are a lot of different disciplines within the employee ranks in a utility. They all derive a way of thinking from their training, and sometimes this coordination becomes difficult. There are people trained in different ways who are somehow trying to communicate with each other.

The employees form a bureaucratic staff. For some reason, the term "bureaucracy" is usually reserved for government and is used with a sense of opprobrium. The fact remains that any business or large-scale enterprise cannot be operated without a bureaucratic staff.

A bureaucracy is organized on the principle of hierarchy, but the mode of orders or directions is based on rationality. Thus the guiding pursuit is to find the rational, the objective, the factual solution, although the solution will be expressed in directives and policies issued from the highest levels of the hierarchy and binding on all below. Nonetheless, despite the authoritative quality of the order, an irrational one is not likely to be obeyed—or at least it will be disputed. Thus, the search for the rational persists.

Therefore, a bureaucracy is not only necessary, but also efficient and appropriate. In fact, the famous German sociologist Max Weber characterizes the bureaucracy as "the most rational known means of carrying out imperative control over human beings."[3]

On the other hand, rationalities harden into routines and procedures, until finally the system becomes more important than the issues. A constant reviving of rationality and a quest for understanding the real nature of the problems is an incessant duty of management.

All of the skills must be constantly directed toward the objective of an enterprise which is basically a technical one: how to obtain and deliver a utility service at the lowest possible cost, yet recognize within the costs all the related social and environmental factors. The bureaucratic staff must be pushed toward these considerations.

In an unorganized group, the employees can use their skills to seek employment elsewhere if they feel their compensation is not adequate or if they are not receiving due consideration of their point of view. In the organized group, such as a union, the employees have the benefit of collective bargaining process, which can result in a strike as the last resort.

Thus, employees are the social entity which is the corporation. The level of their competence, the quality of their morale, and the effectiveness of the leadership they receive are clearly the criteria which determine the success of the entire operation.

The Community

Finally, the utility interfaces with the community. The community is comprised of customers who fulfill many other roles and develop viewpoints and interests from all the roles they fulfill. It is in the aggregate of their composite roles that they express the opinions of the community. Beyond the local community are the larger communities which are regional, national, and, indeed, worldwide. From all these communities come the stream of social, economic, and political forces which have an impact on all their entities and industries.

The agency which expresses the community is the government, which, unfortunately, in popular conception is often equated with the community. Government is only one of the agencies of the community—not inferior to the others, but not always superior. Government, too, is comprised of agencies which in turn are comprised of people operating within the institutional patterns of behavior which their roles assign them. But merely because they have assigned government roles and are endowed with semantic indexes of superiority, their governmental roles, out of the raw fact of being governmental, do not supply them with any wisdom, capabilities, or understanding of the public interest superior to that of their "private" counterparts.

Governmental agencies, however, operate through law, and in those areas where law has the capacity to be effective (e.g., criminal law), it is the best way to get things done. But law has its limitations, and it appears that we have been through a cycle, where the domination of law and legal processes has been too great. A lessening of the impositions could be very beneficial.

Nonetheless, a utility company, perhaps like no other entity, is deeply engaged in interaction with governmental agencies. Among the federal agencies with which it deals are the Federal Energy Regulatory Commission, the Securities and Exchange Commission, the Environmental Protection Agency, the Corps of Engineers, the Internal Revenue Service, the Occupational Safety and Health Administration, and agencies dealing with equal employment opportunities and employee welfare plan practices. State agencies with which a utility interacts include the Public Service Commission, the Department of Environmental Conservation, and agencies dealing with taxation. All of this involves a great amount of work. Over long periods of time, it becomes extremely burdensome, especially when several agencies have overlapping rules, functions, and jurisdictions.

The genius of the free enterprise American utility, however, rests precisely in the system of checks and balances which the interaction with the community and government affords. Through this interaction, not directive from the governmental side, the myriad definitions in the particular cases and situations of the public interest is developed.

The Interplay of the Groups
The organizational structure of the utility itself and the behavior which its customs and work patterns require serve as an umbrella under which the investors, customers, employees, and community interact.

When all goes well and there are no abrupt changes in the relationship among the groups, the enterprise operates efficiently and the various groups are satisfied. But if any group begins to believe that it is not receiving its just reward, or there is an interruption in the norm—for example, when fuel oil prices increased abruptly after the oil embargo of 1973 —each of these groups can express its dissatisfaction.

Customers can oppose the company in rate cases, in construction and other development programs, or in other matters before regulatory agencies. They can develop political support and, through the public media and other means, evoke a sympathy for their point of view. Investors can withdraw their support of the company by disposing of their investments, not subscribing to additional securities, or by demanding premium returns for their money. Employees can bring many pressures to bear on their own behalf, such as a strike.

The composite of the operation of the utility company is therefore the stream of the impacts of these individual groups. In the conventional sense, the utility cannot set goals; it cannot reach for greater and greater profits; it cannot aim to serve a larger territory although it will clearly increase in size. Instead, it is best represented by a flow diagram in which the process steadily moves on. The enterprise then succeeds when all of its constituents are reasonably satisfied. And if the flow is smooth, enormous satisfaction can result.

The investor enjoys a steady and reasonably rewarding return; he does not obtain speculative profit or the gains resulting from growth. But his return is steady and regular.

The customer obtains all the technological benefits which flow from the marvelous results of the utility operation, such as light, heat, the saving of labor, and enjoyment of sound and visual effects. He may even benefit from lower rates, or at least ones that are rising proportionately more slowly than those for other goods and services.

The employees enjoy stable careers and the satisfaction of service to the community.

The community benefits by having a steady enterprise which is inextricably bound up in the local community. The utility cannot move elsewhere to take advantage of lower labor costs or taxes. It must operate locally, paying taxes, creating jobs and cash flow, and generating local economic benefits. It supports other local organizations, charitable and otherwise, and helps create a balanced pluralistic community in which its institutional and economic strength serves as a counterbalance to the political mechanisms.

If the company's financial resources are limited, it fulfills all of these functions poorly. If it is ebullient, it has an advantageous economic and social impact on the community. If the enterprise fails because of over-insistent demands from any one group, all suffer.

As public policy seeks to find ways to use the private sector to advance the public good through legislative enactments,[4] the utility company already exists as a structure available for appropriate legislative encouragement to carry out the desired ends.

The role of management is to try to keep all of these interests flowing smoothly. The role of the board of directors should be to evaluate and make those decisions which take into account all of these diverse interests. While the board is the representative of the shareholders, it serves their interest best to the degree that it serves the interests of all groups which function in relationship with the company.

To achieve and maintain the optimum benefits for all of those interests, the utility must be constantly striving toward an equilibrium, a balance of all competing considerations. The utility is the interface between management and labor, between customers and investors. In responding to this interaction, the utility company, like many other organizations, represents the pluralism which is characteristic of the American system. Through its operation in that pluralism, it provides not only a large share of our material comfort, but also contributes to establishing those life values such as freedom and equity which we deem most important.

Chapter 5

THE TECHNOLOGICAL AND SOCIAL BENEFITS GENERATED BY A UTILITY

The process described in the previous chapter has created an industry with seven basic positive attributes which are, necessarily, characteristic of each company. They are:

1. *The industry's extraordinary technical competence and pragmatic business know-how is the result of generations of growth and development.* Made up of an aggregate of private utilities and municipal and public systems of several kinds, the industry provides electric service throughout the country. This service entails the building of plants for the generation of electricity, the transmission of electricity through transmission systems, and the distribution of electricity throughout the country to the individual retail users.

At present, there is scarcely a part of the country which cannot be reached by electric lines, and almost no one needs to be without electricity if he wants it. This accomplishment of providing electric service throughout the country is taken for granted, but it requires a complex system.

To produce this system and to enable it to work daily is the result of a long period of devoted effort by innumerable people, ingenious invention, and individual mechanical skills.

The industry's infancy is well-known. All of it was very dramatic and has been recorded many times. The story of Thomas Edison's invention of the electric light is legendary, but his subsequent work on developing systems for supplying electricity, though far less famous, was equally dramatic. Edison's work on the development of generators and distribution systems, and the development of electric light companies to provide and carry out this service, was every bit as significant as the light bulb.

Parallel with his work on the incandescent light bulb was the design of a system which could make the light practical and usable. This system included construction of a distribution network, the improvement of the generator, invention of a durable, high-resistance light, development of devices for voltage control and safety (such as fuses and insulation materials), light sockets with switches, and the use of parallel circuits. All of this was the foundation of the entire industry, which began with New York City's Pearl Street Station, the first system for providing incandescent lighting.

Edison's work was followed by that of a host of others, both in the technical and entrepreneurial areas. Companies started up all over the country to manufacture electrical equipment and develop generation and distribution.

Among the portentous developments were the dramatic efforts to utilize the power of Niagara Falls. After much investigation and consideration of competitive ideas, an alternating current electrical system was decided upon and became a sensational reality. With this, central station generation with local distribution became the pattern for national electrification.

Such an ambitious undertaking required the development of innumerable companies to manufacture electrical equipment. The entrepreneurial opportunists did not wait for the development of a master plan, but instead drew upon the energies and imaginations of vast numbers of people throughout the country.

Local electric companies developed everywhere. As the size of generators increased and the economies of scale became evident, consolidations of small companies into larger ones were concluded, until the vast industry that functions today with unusual efficiency was developed. This industry is characterized by both a uniformity and a diversity. There are small companies as well as large ones.

This development of the electric light system, from its infant operation to the complex system of today, required a long history of growth in many areas. The public often does not realize how large and complex a modern power plant is. It took an incredible amount of ingenuity to develop the technology that makes a power plant work. Developing such techniques required not only the basic scientific idea, but also the production of blueprints and drawings and architectural planning. This, in turn, required an entire generation of electrical engineers and others who were trained in the theoretical elements of the process and then spent lifetimes in the actual development of the machinery itself.

The technology of power plants is truly one of components. One usually just thinks of the larger systems such as the generator and the boiler, but all the little links are necessary, too, in order to make the system work. It then becomes a matter of incorporating the intricate science of metallurgy with the use of materials for pipes, tubes, bands, and other devices, the development of metals which resist heat and won't corrode, and the mechanics of fuel burning and ash and waste disposal.

Even within the last 20 or 30 years, plants have grown from 60 megawatts to 1,200 megawatts in size, and all of this growth has required use of new techniques.

The science of transmission has gone through the same kind of evolution and developing complexity as the generation process. The science of transmission has grown vastly from the early days when central stations had to be near load centers. Now they may be miles and miles away.

Transmission first involved relatively small alternating current lines. Eventually these were replaced by the enormous direct current lines which are built today. And the transmission of power is becoming more complex than ever because of the move to higher voltages, such as 765-kilovolt lines.

New York State is now working on a direct line with Canada which will permit both Ontario Hydro and Quebec Hydro to transmit higher-voltage power from the upper reaches of Canada to New York City. This is possible only with the very complex development of high-voltage transmission techniques.

In the same way that the transmission lines were developed, the distribution system, which used to consist of very primitive lining of wires, is now a very complex system. And the technological problems are huge indeed, especially in a city like New York where the whole system has to be underground.

With the development of the understanding of atomic power during World War II, and with the Atoms for Peace Program of the Eisenhower administration, the electric industry experimented with nuclear power. In terms of the problems involved and the scientific issues that had to be resolved, the nuclear industry is unbelievably complex. Characteristic of any novel and growing industry, many problems still require resolution and understanding.

All of these things were done under the aegis of a free enterprise system. While other countries in the world without such a system have been able to duplicate the technology of the American system, by and large the pioneer work—the whole concept of extending electricity to everybody—was started and moved forward by the United States.

The impressiveness of the system is dramatized by comparing it with the infrastructure, consisting of highways, bridges, tunnels, water supply and sewer supply systems and mass transportation. In some cases, mass transportation systems have almost broken down and important and impressive bridges like the famous Brooklyn Bridge are reported to be in deteriorating condition.

The dreadful situation is the result of years of neglect and improper planning by public officials who have not adequately provided maintenance funds for these structures. Even worse, in many cases, funds marked for maintenance have been siphoned off to pay for deficits in operating budgets.

Many estimates as to the needs have been developed, including a recent one for the New York Tri-State area, for which a recent study of the Regional Plan Association estimates that $8 to $9 billion a year will be needed for the rest of the current decade to reclaim and expand the so-called infrastructure.

This incredible need arises at a time when governmental finance is in dreadful condition and state and local governments are bordering on bankruptcy, or, in effect, are in bankruptcy although it is not designated as such.

Capital assets in the public sector have been physically deteriorating without any provision for depreciation accounting and without any accumulation of funds for depreciation reserves.

In contrast, the utility process has provided for adequate flows of revenue and appropriate accounting including depreciation accounting with the result that the systems are in a constant state of rehabilitation. Among other things, as previously mentioned, the mortgage bond indentures of utilities require annual inspections by independent engineers and the supplying by them of certificates that the systems are in good order and working condition.

Utility rates are complained about but at least they are rational and responsible and provide the means whereby the sophisticated technical equipment, which is necessary to supply electricity, is constantly being modernized and up-dated. The industry indeed represents a huge system which confers great technological benefit on the public.

2. *In addition to its technological competence, the utility industry has the form of corporate organization which brings together the ingredients of management, capital, and labor which are essential to providing the useful product and service it supplies.* The industry represents the unique free enterprise style which has brought benefits literally beyond parallel in the history of our civilization. The organizations which the public utility represents provide the potential for the industry to further enhance human fulfillment. The details of this thought are expressed throughout this study. Without amplification here, point 2 is stated as one of the benefits rendered by the utility corporation.

3. *As a consequence of these technical skills and organizational attributes, the industry has achieved an extraordinary record of reliability of service.* This remarkable performance is true throughout the country and is not matched anywhere in the world. In 1978 in New York State, for example, electric customers received electric service 99.0865 percent of the time. Where there were interruptions, the average duration per interruption was only 1.18 minutes. The number of interruptions per 10,000 customers was 10.13.

While this outstanding performance may be taken for granted, it actually requires a high level of expertise and competence to achieve. It can only be appreciated by someone who has lived in other countries where interruptions, failure of service, and rotation of blackouts are common occurrences.

4. *Utility companies pay enormous taxes, which make them a significant factor in enabling government programs to function.* It is appropriate that utility companies and utility transactions be taxed because no economic activity should take place without taxes being paid. The industry, nevertheless, questions whether the utility companies and their customers are paying more than their fair share of taxes. They are making a concerted protest on this issue to state and federal legislators. The industry argues that public bodies, often referred to as public benefit corporations, bestow

little or no benefit upon the public through the payment of taxes.

Tables A, B, and C show the taxes paid by seven gas and electric corporations in New York State for the years 1980, 1981, and 1982, respectively. They show that the grand total was $1.56 billion in 1980, $1.72 billion in 1981, and $1.84 billion in 1982.

The tables do not show the taxes based on income received from utility companies. That question involves so many unknowns that a figure cannot be reliably supplied. However, these taxes, just like the taxes paid by employees, are generated out of the business. In a rate case, the rate of return authorized is a rate before taxes to the investor. Accordingly, the taxes paid by the employees are a component of the total taxes paid which are generated out of the enterprise.

The taxes paid by the utility also clearly demonstrate the flow characteristic—the cash flow, the tax flow, etc.—of the utility activity which has been alluded to frequently in previous sections. Table D traces the taxes referred to in Tables A, B, and C.

In conclusion, the utility process provides a very fertile source of tax revenue and makes an appropriate contribution to the public welfare. The tax bill represents the social cost of doing business and illustrates one of the prime attributes of the private-sector operation; namely, it provides a social means of allocating the full cost of a product or service, while a so-called public entity tends to charge an imaginary too-low cost that neglects the social cost component of the public welfare.

5. *Utility companies have created a vast number of jobs and presently employ about 500,000 people nationally.* They also have the capacity to create more jobs and more prosperity, if certain political issues could be resolved.

6. *Utility company securities are fundamentally important sources of income for many individual investors (many of whom are far from affluent and depend on such income for some of their support).* For example, a recent sample of 11 companies' shareholder surveys "shows 51 percent of these shareholders to be retirees. Over half of them are over 65 years of age. . . . About one-quarter of them have an annual household income of $15,000 or less. . . ."[1]

Utilities are an integral part of the portfolios of pension and welfare funds, insurance companies, mutual funds, and others. The steady flow of income generated by public utility operations through yields on investments is basic to the structure of our national economy.

With the number of people who own utility stocks and depend on them for a source of income, the steady flow of dividends and bond payments made by utility companies has been a major factor in the income flow of the country.

But if utilities were to develop severe financial problems which led to bankruptcy, there could be economic Three Mile Islands (Three Mile Island was not only an engineering problem but also, as people in the industry so well know, a financial crisis.)

If a situation which interferred with a utility's ability to pay dividends

Table A
The Energy Association of New York State[1]
Tax Survey—1980

	Central Hudson	Consolidated Edison	Long Island Lighting
LOCAL TAXES			
Local Taxes Paid By Utility			
1. Real Property			
a. Real Estate	$ 7,533,028	$192,994,127	$ 88,871,288
b. Special Franchise	4,164,681	182,426,064	43,733,561
2. Gross Receipts			
a. Utility Services			
(Municipal Gross Income)	590,849	78,269,401	3,438,936
b. Special Franchise			
(New York City Only)	0	5,726,143	0
3. Sales and Use			
(Company Purchases)	268,164	43,748,422	2,181,713
4. Occupancy & Commercial Rent	2,006	405,891	1,305
5. Motor Vehicle	0	116,313	275
6. Other	0	* 26,774	0
Sub Total	$12,558,728	$503,713,135	$138,227,078
Local Taxes Billed Customers			
1. Sales and Use	2,739,964	112,886,000	24,072,732
TOTAL—LOCAL TAXES	$15,298,692	$616,599,135	$162,299,810
STATE TAXES			
State Taxes Paid By Utility			
1. Gross Receipts			
a. Utility Service			
(State Gross Income)	7,872,473	107,309,721	36,737,432
b. Franchise	2,257,813	28,816,045	9,212,099
2. Excess Dividends	444,770	5,666,340	4,384,323
3. Sales and Use			
(Company Purchases)	581,157	8,995,669	2,830,278
4. Unemployment	129,396	2,140,835	569,108
5. Truck Mileage	13,261	91,606	15,370
6. Mortgage	479,718	0	519,874
7. Organization	0	0	0
8. Vehicle Registration Fees	35,715	225,000	189,418
9. Gas & Diesel Motor Fuel	35,527	315,229	207,919
10. Disability Benefit Contribution	0	38,538	8,331
11. Other	0	**** 904	0
Sub Total	$11,849,830	$153,599,887	$ 54,674,152
State Taxes Billed Customers			
1. Sales and Use	4,420,052	90,889,000	25,839,374
TOTAL—STATE TAXES	$16,269,882	$244,488,887	$ 80,513,526
TOTAL LOCAL AND STATE TAXES PAID BY UTILITY AND BILLED CUSTOMERS	$31,568,574	$861,088,022	$242,813,336

*Leaded Gasoline and Real Property Transfer Taxes   **Mortgage Recording Tax
***Water Pollution Control   ****Real Property Transfer Tax

[1]An unincorporated association whose members are: Central Hudson Gas and Electric Corp., Consolidated Edison Company of New York, Inc., Long Lighting Company, New York State Electric & Gas Corp., Niagara Mohawk Power Corp., Orange and Rockland Utilities, Inc., and Rochester Gas and Electric Corp.

N.Y. State Electric & Gas	Niagara Mohawk	Orange and Rockland	Rochester Gas & Electric	TOTAL
$19,964,136	$ 71,278,490	$11,923,955	$18,448,596	$ 411,013,620
9,833,082	35,425,231	6,094,445	10,402,758	292,079,822
2,472,764	10,005,183	721,159	5,174,942	100,673,234
0	0	0	0	5,726,143
877,572	2,473,178	2,858	1,012,721	50,564,628
0	2,809	0	0	412,011
0	0	0	0	116,588
0	** 1,963	0	*** 435,974	464,711
$33,147,554	$119,186,854	$18,742,417	$35,474,991	$ 861,050,757
9,103,265	21,964,443	1,129,046	5,210,122	177,105,572
$42,250,819	$141,151,297	$19,871,463	$40,685,113	$1,038,156,329
19,095,384	48,237,761	7,616,595	13,386,771	240,256,137
5,176,235	12,733,660	2,045,073	3,513,096	63,754,021
1,691,685	3,627,817	471,607	1,008,454	17,294,996
1,593,001	4,306,661	624,182	1,350,295	20,281,243
537,203	1,182,934	200,749	253,708	5,013,933
41,611	75,365	12,205	10,212	259,630
0	734,003	0	543,825	2,277,420
0	0	0	0	0
146,551	(Est.) 282,169	48,225	44,885	971,963
182,600	(Est.) 393,800	69,582	68,036	1,272,693
0	0	0	4,450	51,319
0	0	0	0	904
$28,464,270	$ 71,574,170	$11,088,218	$20,183,732	$ 351,434,259
11,522,717	24,291,905	4,839,804	7,667,839	169,470,691
$39,986,987	$ 95,866,075	$15,928,022	$27,851,571	$ 520,904,950
$82,237,806	$237,017,372	$35,799,485	$68,536,684	$1,559,061,279

Note: Of each retail customer dollar received by the companies, 19 cents are paid as taxes or fees to state and local governments.

Table B
The Energy Association of New York State[1]
Tax Survey—1981

	Central Hudson	Consolidated Edison	Long Island Lighting
LOCAL TAXES			
Local Taxes Paid By Utility			
1. Real Property			
a. Real Estate	$ 8,586,230	$200,918,368	$100,159,558
b. Special Franchise	4,788,866	198,750,078	48,601,864
2. Gross Receipts			
a. Utility Service (Municipal Gross Income)	702,464	99,236,723	4,492,452
b. Special Franchise (New York City Only)	0	7,080,796	0
3. Sales and Use (Company Purchases)	291,883	53,370,519	2,229,679
4. Occupancy & Commercial Rent	2,218	680,893	35
5. Motor Vehicle	0	114,654	462
6. Other	0	* 19,136	0
Sub Total	$14,371,661	$560,171,167	$155,484,050
Local Taxes Billed Customers			
1. Sales and Use	3,497,216	143,466,147	22,961,786
TOTAL—LOCAL TAXES	$17,868,877	$703,637,314	$178,445,836
STATE TAXES			
State Taxes Paid By Utility			
1. Gross Receipts			
a. Utility Service (State Gross Income)	9,819,064	134,220,951	48,038,347
b. Franchise	2,851,549	36,018,693	11,570,053
2. Excess Dividends	577,620	6,274,294	5,561,625
3. Sales and Use (Company Purchases)	594,969	11,523,291	3,004,712
4. Unemployment	119,726	2,106,773	551,980
5. Truck Mileage	13,981	69,870	14,162
6. Mortgage	287,796	0	3,112,718
7. Organization	0	0	0
8. Vehicle Registration Fees	38,050	222,067	186,377
9. Gas & Diesel Motor Fuel	44,421	298,424	213,202
10. Disability Benefit Contribution	0	40,122	4,856
11. Other	0	**** 181	0
Sub Total	$14,347,176	$190,774,666	$ 72,258,032
State Taxes Billed Customers			
1. Sales and Use	2,747,994	73,405,946	18,014,318
TOTAL—STATE TAXES	$17,095,170	$264,180,612	$ 90,272,350
TOTAL LOCAL AND STATE TAXES PAID BY UTILITY AND BILLED CUSTOMERS	$34,964,047	$967,817,926	$268,718,186

*Leaded Gasoline and Real Property Transfer Taxes   **Mortgage Recording Tax
***Water Pollution Control   ****Real Property Transfer Tax   †New York State Excise Tax

[1]An unincorporated association whose members are: Central Hudson Gas and Electric Corp., Consolidated Edison Company of New York, Inc., Long Island Lighting Company, New York State Electric & Gas Corp., Niagara Mohawk Power Corp., Orange and Rockland Utilities, Inc., and Rochester Gas and Electric Corp.

N.Y. State Electric & Gas	Niagara Mohawk	Orange and Rockland	Rochester Gas & Electric	TOTAL
$22,890,605	$ 75,661,117	$13,778,033	$21,382,000	$443,375,911
10,772,050	42,707,005	7,033,937	11,593,714	324,247,514
2,844,869	10,969,411	863,839	7,098,386	126,208,144
0	0	0	0	7,080,796
1,654,040	2,663,623	2,332	1,144,917	61,356,993
0	2,766	0	0	685,912
0	0	0	0	115,116
0	** 2,840	0	*** 438,761	460,737
$38,161,564	$132,006,762	$21,678,141	$41,657,778	$ 963,531,123
9,101,954	21,832,561	1,019,207	3,452,356	205,331,227
$47,263,518	$153,839,323	$22,697,348	$45,110,134	$1,168,862,350
22,709,639	58,019,374	8,979,797	16,524,279	298,311,451
5,761,775	15,364,287	2,407,279	4,316,981	78,290,617
2,006,529	4,182,480	521,502	1,234,959	20,359,009
2,482,782	4,431,212	870,145	1,526,556	24,433,667
505,091	1,024,233	181,840	260,348	4,749,991
39,934	83,617	7,493	9,376	238,433
1,051,299	922,992	0	490,779	5,865,584
0	0	0	0	0
157,043	(Est.) 291,954	44,017	36,769	976,277
171,564	(Est.) 435,800	69,459	68,675	1,301,545
0	0	0	4,813	49,791
0	0	0	† 371	552
$34,885,656	$84,755,949	$13,081,532	$24,473,906	$ 434,576,917
5,364,217	13,292,438	1,983,980	3,978,809	118,787,702
$40,249,873	$98,048,387	$15,065,512	$28,452,715	$ 553,364,619
$87,513,391	$251,887,710	$37,762,860	$73,562,849	$1,722,226,969

Note: Of each retail customer dollar received by the companies, 17 cents are paid as taxes or fees to state and local governments.

Table C
The Energy Association of New York State[a]
Tax Survey—1982

	Central Hudson	Consolidated Edison	Long Island Lighting
LOCAL TAXES			
Local Taxes Paid By Utility			
1. Real Property			
a. Real Estate	$ 9,180,999	$218,293,674	$111,937,070
b. Special Franchise	5,061,412	217,016,199	51,038,132
2. Gross Receipts			
a. Utility Service			
(Municipal Gross Income)	797,147	103,345,984	4,569,825
b. Special Franchise			
(New York City Only)	0	7,125,439	0
3. Sales and Use			
(Company Purchases)	259,647	48,669,673	2,725,983
4. Occupancy & Commercial Rent	2,228	326,978	0
5. Motor Vehicle	0	110,645	520
6. Other	0	[b,e] 67,524	0
Sub Total	$15,301,433	$594,956,116	$170,271,530
Local Taxes Billed Customers			
1. Sales and Use	2,836,495	144,141,912	17,607,525
TOTAL—LOCAL TAXES	$18,137,928	$739,098,028	$187,879,055
STATE TAXES[h]			
State Taxes Paid By Utility			
1. Gross Receipts			
a. Utility Service			
(State Gross Income)	10,564,646	140,402,644	47,869,363
b. Franchise	2,992,521	37,027,142	11,625,974
2. Excess Dividends	671,238	7,657,884	7,239,800
3. Sales and Use			
(Company Purchases)	704,445	10,489,918	3,692,993
4. Unemployment	128,000	2,134,267	556,441
5. Truck Mileage	15,315	62,710	12,243
6. Mortgage	430,749	0	2,173,405
7. Organization	0	0	0
8. Vehicle Registration Fees	37,935	216,812	187,874
9. Gas & Diesel Motor Fuel	48,237	234,288	181,355
10. Disability Benefit Contribution	0	35,338	5,219
11. Other	[g] 65	[e,g] 34,878	0
Sub Total	$15,593,151	$198,295,881	$ 73,544,667
State Taxes Billed Customers			
1. Sales and Use	2,859,651	81,555,759	18,020,310
TOTAL—STATE TAXES	$18,452,802	$279,851,640	$ 91,564,977
TOTAL LOCAL AND STATE TAXES PAID BY UTILITY AND BILLED CUSTOMERS	$36,590,730	$1,018,949,668	$279,444,032

[a]An unincorporated association whose members are: Central Hudson Gas & Electric Corp., Consolidated Edison Company of New York, Inc., Long Island Lighting Company, New York State Electric & Gas Corp., Niagara Mohawk Power Corp., Orange and Rockland Utilities, Inc., and Rochester Gas and Electric Corp.

[b]Leaded Gasoline   [c]Mortgage Recording Tax   [d]Water Pollution Control
[e]Real Property Transfer Tax   [f]New York State Excise Tax   [g]Hazardous Waste Tax
[h]The 1982 MTA Business Tax Surcharge is not included in this survey since it is being included in customer rates in 1983.

N.Y. State Electric & Gas	Niagara Mohawk	Orange and Rockland	Rochester Gas & Electric	TOTAL
$27,287,965	$ 88,810,359	$14,397,285	$21,686,060	$491,593,412
11,145,789	40,953,075	7,780,878	12,835,172	345,830,657
3,286,076	12,469,116	1,040,916	9,210,093	134,719,157
0	0	0	0	7,125,439
1,415,344	2,956,624	3,180	1,297,787	57,328,238
0	3,620	0	0	332,826
0	0	0	0	111,165
0	(c) 8,054	0	(d) 995,118	1,070,696
$43,135,174	$145,200,848	$23,222,259	$46,024,230	$1,038,111,590
10,806,972	22,907,162	1,254,898	3,926,171	203,481,135
$53,942,146	$168,108,010	$24,477,157	$49,950,401	$1,241,592,725
26,264,438	64,544,309	10,680,061	19,107,851	319,433,312
6,565,214	17,066,253	2,785,623	4,864,191	82,926,918
2,801,949	5,930,251	553,549	1,216,221	26,070,892
2,235,102	5,003,609	763,516	1,730,383	24,619,966
460,547	955,323	163,130	265,899	4,663,607
42,021	79,088	7,712	9,871	228,960
1,576,187	3,053,680	0	593,880	7,827,901
0	0	0	0	0
172,124	297,177	44,645	40,532	997,099
167,531	410,400	61,775	59,343	1,162,929
0	0	0	4,341	44,898
0	(g) 1,752	0	(f,g) 1,722	38,417
$40,285,113	$ 97,341,842	$15,060,011	$27,894,234	$ 468,014,899
6,352,647	15,498,081	2,406,184	4,527,000	131,219,632
$46,637,760	$112,839,923	$17,466,195	$32,421,234	$ 599,234,531
$100,579,906	$280,947,933	$41,943,352	$82,371,635	$1,840,827,256

Note: Of each retail customer dollar received by the companies, 17 cents are paid as taxes or fees to state and local governments.

Table D

The Activity	The Tax
1. The investor supplies capital which is used to purchase property on which there are assessed . . .	property taxes.
2. The investor supplies capital which provides for construction of the plant . . .	During the construction period, the constructors pay out of the utility capital payments the various kinds of taxes for which they are liable.
3. The construction paid for by the investor funds increases the value of the property and supplies increments to . . .	property taxes.
4. When the plant becomes operational, electricity is sold to customers upon which is imposed . . .	sales taxes.
5. The revenue received by the utility from its customers is subject to . . .	revenue taxes.
6. The revenue received by the utility is used to pay (a) for fuel and supplies . . .,	the prices of which, in turn, reflect all the taxes paid by the suppliers;
(b) wages and salaries on which are paid . . .	payroll taxes, on which the recipients pay income taxes;
(c) interest on debt on which recipients pay . . .	income taxes.
7. After providing for all expenses, the revenue results in income, on which the utility pays . . .	income tax.
8. After providing for all expenses, including taxes, the utility pays dividends to shareholders, on which they pay . . .	income taxes.

46

and interest developed, there would be an unbelievable impact on the national economy.

7. *Because of their service-oriented nature, utility companies and their employees, perhaps more so than in other industries, contribute in an infinite number of ways to the communities in which they are located.* For example, they support community and civic activities, such as the United Way and hospital and educational institutions. Using the resources at their command, they respond in emergency and disaster situations, such as snowstorms and hurricanes.

Furthermore, utility companies are inextricably bound to the communities they serve. They cannot move from where they are. If they are located in the East, Northeast, or Middle West, they present no threat that they will move to the South, Southwest, or West, as other businesses do. In short, they provide a solid, dependable core of employment, and have employees who become an integral part of the communities where they work and live.

In addition, utilities are stabilizing influences in the communities. They are not boom or bust, they don't expand rapidly and attract employees or other businesses to the area in large numbers and then lay the employees off or cut back in other ways. Instead, utilities represent stable, solid enterprises which provide steady, solid growth.

Chapter 6

THE PRIVATE SECTOR UTILITY COMPARED WITH THE PUBLIC SECTOR UTILITY

The issue of public versus private power is engulfed in a great deal of emotional reaction and naiveté. There is a mystique about public power, and a religious fervor about the preference clause and other procedures which express the same idea. At the same time, there are complaints about private sector profits and claims of utility insensitivity to public concern, all coupled with expectations of a panacea under a public power operation.

The issues, of course, are very complex, and should not be clouded by color words or inflammatory slogans. Instead they must be viewed in terms of actual processes.

The basic issue, one of policy and long-term objectives, is also related to fundamental distinctions between economic and political processes. It is also necessary to be aware of a basic distinction in public power operations.

Public power agencies selling at retail and those selling at wholesale encounter vastly different pressures. Wholesalers, by and large, can support their operations by contracts with other entities, both public and private, and are assured a financial stability through legal commitments. Indeed, their bond financings are only possible because of the revenue assurance which these contracts provide. Furthermore, such entities are insulated from the public because wholesalers are not selling to the public. When wholesalers are formed as separate entities, they are usually organized in some form of public authority or public corporation. In many ways, they are not unlike private utilities. And to the extent that they operate as economic entities rather than political ones, they avoid many of the inadequacies of a strictly governmental operation.

However, public agencies that operate as part of a governmental unit, such as a municipality, sell directly to the public, and are not separately incorporated invite a whole series of vices. In short, when public agencies operate as political entities rather than economic ones, the difficulties become manifold.

Underlying all of these problems is the distinction between economic and political processes. Where economic activity is determined by the economic processes—clearly the case of a private utility—the long-term results are beneficial; where the economic process is operated as a political process, or as part of the political process, disaster is almost inevitable.

The economic process, which is concerned with accumulating capital and using it in production and distribution to produce a financial return, seeks stability and the certainty which will enable the production and distribution to take place. The process demands continuity of personnel and the development of expertise. Its managers remain in control as long as they manage the enterprise to produce an appropriate return. In the long run, the economic process operates for the benefit of the community at large by providing the goods and services which the populace needs or desires.

The political process, on the other hand, is constrained by the need to respond to periodic elections. It too seeks to find stability, but its practitioners must always live in an atmosphere of instability and in the shadow of elections every two or four years. They must appear to be doing those things which favor the populace and make their lives easier and more convenient.

A utility company, however, is totally an economic process and operates as an independent entity which must provide for its subsistence solely from the revenue flow from its customers. It is a self-sufficient budgetary unit, and the need to maintain the budgetary integrity of a private sector entity provides a powerful social mechanism dictating corporate action. This may not be popular or easy, but it is necessary if the integrity is to be maintained. A corporate officer must operate his business in such a fashion that it yields a return; otherwise, he is subject to invective and professional disrespect, a decrease in the value of any equity ownership which he might possess, possible loss of his job, and possibly suits by stockholders and other investors. Thus, while he does not want to raise the prices charged for his company's product and is disturbed by public and media criticism of any price increase, he faces this unpleasantness rather than injure the financial integrity of his enterprise.

On the other hand, the finances of the public sector agency, which is not formed as a separate authority, become part of the general governmental budget. As long as the power to tax remains, the budgetary deficits of the entity can be covered up by subsidies, transfer of assets, or avoidance of legitimate charges against revenues, such as depreciation. In this way, the bitter necessity of a rate increase is avoided. When the corporate authority stems from the political systems, when the vote of the electorate can be determined, the opportunity to avoid imposing a definable, noticeable, regularly payable increase is tempting indeed; resorting to the general tax fund to defray increasing costs seems more appropriate.

The process is illustrated very vividly by the English electric utility system. In England, electric rates must not only go through a technical and accounting scrutiny, but also must be approved by a political organization of the government. The government has regularly denied rate increases for social and political reasons, and any considerable deficits have been absorbed through tax subsidies. As a result, the English utility system is far from self-supporting and is a burden upon the taxpaying public.

The same process has occurred repeatedly with mass public transportation in this country. The New York City subway system is probably the outstanding example. Here, a justified increase in rates is avoided periodically. Because the full cost of service isn't raised from riders and tax subsidies have become more and more difficult to obtain, the operation and maintenance of the system has been neglected until it is nearly in a state of collapse.

But this is only part of the issue. The fact that several public entities have performed poorly does not detract from the fact that many of them have performed very well.

Indeed, while the electric utility responsibility in the United States has been basically discharged by the private sector, there are substantial public sector utilities. These include federal agencies like the Tennessee Valley Authority and state agencies like the New York State Power Authority, both of which were formed as quasi-economic entities. There are also a host of municipal systems and federally financed local entities, such as the rural electric cooperatives.

While some municipal operations are very poor, many of these governmental agencies have performed very well indeed. Their operations are technically sound, they provide good service, and they are staffed with competent personnel who maintain high standards of performance. In this perspective, it is difficult to argue that the private sector companies do a better technological job than the public sector entities.

In terms of long-term national goals, however, the private sector is much preferred. On the surface, the public sector utility appears attractive. It provides low rates and its designation as "public" seems to create a confidence and an immunity from the alleged vices of private profit. But in terms of the total picture, the public entity clearly provides less public benefit than the private sector utility. Inherent in the public sector utility's mode of operation are many contradictions and disabilities.

For example, the Power Authority of the State of New York was established in 1931 to meet some very special problems. It constructed electric generating facilities by the International Rapids section of the St. Lawrence River. The Authority remained relatively inactive until 1951, when the question arose about the appropriate development of the Niagara Falls' power potential. A proposal was created by a combination of New York private sector companies, and a very active political debate ensued whether the Falls should be developed by the Power Authority or by the private companies.

The private company approach was very persuasive, and the principal argument in favor of the Power Authority arose out of the cliché that the public waters should be developed by the people. Because of the apparent political appeal of this fallacious argument, some leading political figures seemed to favor private development. The majority, however, leaned toward using the public power authority. Even so, the private company proposal to designate a private sector company to develop Niagara Falls almost obtained the appropriate federal approval, which was required under the treaty between the United States and Canada.

The situation changed radically, however, when a landslide destroyed Niagara Mohawk Power Corporation's Schoellkopf Plant. As a result of this disaster, Niagara Mohawk felt it could no longer wait for the outcome of a drawnout political struggle, and shifted its support from private sector development to one by the Power Authority. The decision carried the understanding that Niagara Mohawk would receive an untouchable block of power, called replacement power, to make up for the output of the Schoellkopf Plant. In addition, specified blocks of power were designated for high-load factor customers in the northern part of the state, because the Power Authority power was available only in upstate areas. With these steps, the Authority became the major producer of hydropower in New York State.

A little more than a decade ago, Governor Nelson A. Rockefeller established a committee of recognized New York State leaders. The committee included people with recognized expertise in the utility business, both public and private, and prominent citizens with recognized social and political concerns. The committee made a thorough study of the state's power policy and prepared a comprehensive report, called the Folsom Report after the name of the committee chairman.

This report declared that the private power sector should be the basic provider of energy, supplemented from time to time by resources of the Power Authority. This conclusion confirmed a policy which has been effective in New York State since it first had electric power.

In pursuing this policy, the Power Authority admirably has supplied the supplemental power to the private sector. It is a dynamic agency which, in the future, will undoubtedly have a prominent role in undertaking those activities which, for reasons pragmatic, political, and otherwise, appropriately fall within its assigned mission.

Further, in pursuit of this policy, statutes were amended in 1968 to permit the Authority to build pumped storage plants and a baseload nuclear plant, the FitzPatrick Plant.

As a result of the financial crisis at Consolidated Edison, the Power Authority Act was amended again in 1974 to permit the Power Authority to purchase two of Con Edison's plants and to take over the power supply of several public agencies, including the Municipal Transit Authority in New York City. These legislative developments provided a pattern for continual growth of the New York Power Authority as it supplemented private sector development.

During this period, the private sector companies were also growing and being accepted by the public. But as a result of the sharp increase in electric power rates after the 1973 oil embargo, the private sector utilities have recently had a very difficult time. As a result, a renewed interest in expanding the Power Authority has emerged. But when its role tends to become dominant rather than supplemental, all of the dilemmas and public interest drawbacks which are inherent in public power become crucial.

The Power Authority is often thought of as a governmental function, but this is clearly a misapprehension. In fact, when the Power Authority is viewed functionally, it looks like a private utility. All of the nomenclature which describes the Authority as governmental is misleading. The function of government is to adopt laws for the public good, to enforce laws, and to regulate. The Power Authority does none of these. As formed by the state legislature under a statute, it is called a public benefit corporation. It derives all of its powers from that statute.

The New York State utility companies, on the other hand, are formed under the Transportation Corporations Law which, apart from its unfortunate terminology, is a statute that confers upon each utility the rights, privileges, immunities, and powers of a corporation.

Each of these utilities owns property as a separate corporation. The Power Authority, rather than New York State, owns its property. In that respect, it is just like any public utility.

Similar to investor-owned utilities, the funds for the Power Authority are not raised by taxation, but by offering securities in the public markets to private investors. In fact, there is little in these similarities that distinguishes a public entity from the private, and nothing which says that public is better. But there are real differences which tell us that private is better. Among them are:

Capital Structure
The public power agency offers only debt securities; there is no equity component in its financial structure. This is often thought to be an advantage for a public power agency, but it is only an advantage if circumstances favor the operation of the public agency. If the agency's revenues were curbed—and this could happen if the agency has overcapacity or if, for some reason, there were defaults by its customers—the agency would have to default on payment of interest on its bonds, unless some remedy were provided.

The utility company, on the other hand, has the flexibility of equity, and its social function is to provide a margin of safety for bad times. No one expected that this function would come into play, but that is precisely what took place at Consolidated Edison in 1974. When its revenues fell off, Con Edison skipped its dividend instead of going bankrupt. Of course, this created much uneasiness in the financial markets, and it may have been bad for Con Edison's financial reputation. But, interest on bonds continued to be paid and no one filed petitions in bankruptcy courts. All of this leads to speculation that sound financing might well require that the Power Authority have an equity component in the future.

Certainly using the word "public" does not ensure the financial integrity or viability of a public authority. The readjustment in the American economy during the early 1970's brought a number of governmental entities, including some of our major cities to an informal bankruptcy, even though formal court bankruptcy proceedings were not instituted. In New York State, for example, a number of independent authorities—such as the Urban Development Corporation, the Environmental Facilities Corporation, the Housing Finance Agency, and the Medical Care Facilities Finance Agency—had to be rescued from serious financial difficulties by remedial legislation. Financial difficulties are far from unknown to governmental agencies.

Cash Flow

An analysis of the revenue flow of the public power agency and the private utility reveal two striking differences between them.

The first difference—apparent and easy to understand—is that the public agency does not pay taxes and the interest on its bonds is tax-free. (See a below.) The second difference—subtle and yet fundamental to an economic system—is that the public agency does not produce profit, and thus does not contribute to capital formation. (See b.)

(a) Taxes

The failure to pay taxes is not an inherent advantage of a public agency, but is merely a provision of the tax law. What it succeeds in doing is placing the state in the unfortunate position of not utilizing one of the best methods of raising tax money. There is an old adage that a successful sovereign is one who only imposes taxes where taxes are felt the least. From the sovereign's point of view, the taxation of utility companies makes sense because it provides great sources of revenue to the government through taxation of utility revenues. On the other hand, the public agency arrangement is unintelligent because it allows an essential flow of revenue to go untaxed.

The consequence is that the private sector corporation, which is not called a public benefit corporation, conveys great benefit to the public by paying taxes. The taxes paid by combination electric and gas companies to New York State and to its local subdivisions in 1980 were approximately $155 billion. On the other hand, the Power Authority, although called a public benefit corporation, conveys only insignificant benefits through meager payments in lieu of taxes. In fact, the Power Authority and other public agencies are detrimental to the public interest because they siphon off to the public sector revenues which could be taxed.

(b) Capital Formation

All human activity, whether it is primitive or highly sophisticated, should be accompanied by gain. The goal of work is not only to provide maintenance, but also to provide a saving, an increment from the work, which will survive into the future.

The production of electricity is such a monumental exercise of human ingenuity and effort that it should be expected to produce an increment or a profit. The electric industry should be so organized and dedicated to effort and efficiency that it produces a profit. And if our affairs are organized in such a way that all the effort put into the utility industry does not provide a saving, then we indeed are profligate of our national efforts.

Public power just "makes do," but the private sector, on the other hand, produces dividends to the equity holder and an enhancement in values. Out of these profits and value enhancements comes capital formation. For a long time, the country has been very deficient in capital formation. Notwithstanding all of the technical jargon, the basic cause of high interest rates is a shortfall in available liquid capital assets. Of course, adequate equity relief may mean higher rates to customers. But if customers don't contribute such increments to value, they are not paying the full social cost of the service.

In the final analysis, this is the true criticism of the public sector. It does not abstract from the process its full cost, neither in the contribution to taxes nor in its contribution to capital formation.

Responsibility Accounting

In addition, although the Power Authority exercises no governmental function under present statutes, in a sense it has governmental immunity. This immunity isolates the governmental agency from effective control either by the subtle process of social pressures, which requires full disclosure of affairs, or directly through regulation. Social controls of a governmental agency are almost nonexistent or inexpressible and regulation of a governmental agency is inapt, if not impossible.

(a) Absence of Public Disclosure by Government Agencies— Confusion of Terms:

There is very little direct or open disclosure by a governmental agency of its affairs. Indeed, such agencies are protected by the semantics of being public. There is a tacit assumption that since they are characterized as being public, their affairs are in the public domain and no system of disclosure is necessary. Nothing could be further from the truth. Literally, public agencies operate their own affairs unchallenged and unnoticed.

The point is best recognized when the contrast with the private utility is made. The private utility lives in the proverbial fish bowl, its affairs being open and disclosed to the public. Disclosure is required not only by public utility commissions of the individual states and by the Federal Energy Regulatory Commission, but also, most effectively, by the Securities and Exchange Commission through its complex reporting system requiring proxy statements, registration statements, Annual Reports on Form 10K including detailed financial disclosures, Quarterly Reports on Form 10Q, and specific information in Annual Reports to Shareholders. In addition, the so-called "insider rules" make it necessary for any publicly-held corporation to disseminate promptly to the media information about any significant corporate event.

Furthermore, the public agency does not have any constituency to which it must render its accounting. As discussed in Chapter 3, the private corporation is accountable to shareholders, and shareholders perform not only the financial function of providing to the entity the protection of equity, as shareholders, but they also exercise a direct corporate control. While this control is not effectively expressed in voting power, except in the event of proxy contest between competing managements, it is expressed through the function of the annual meeting, with all the subtle social pressures it evokes as described in Chapter 3, and through the continual threat of a shareholder's suit. The ability of the shareholder to respond depends upon his knowledge, and this is supplied in the private sector through the intensive reporting techniques mandated by the SEC.

A convincing close to the discussion is provided by the question: Did you ever hear of an annual meeting of the Power Authority?

(b) Absence of Regulation

It would seem to be a simple solution to the accountability question to suggest that the Power Authority be regulated; but if you provide for regulation of the Authority then a difficult situation arises. For example, if in New York the Public Service Commission were to regulate the Power Authority, what would happen if the Authority objected to a decision of the Commission, or believed that the procedures or process of the Commission were contrary to law? The usual remedy for review of Public Service Commission rulings is an appeal to the courts. If the Power Authority were to initiate such a proceeding, the situation of one state agency suing another would arise. Whether this is legally possible may be debatable; politically, it presents an uncomfortable, if not unsolvable, paradox.

By not substituting a regulated utility for a nonregulated one and by continued reliance on the private sector, the paradox can be avoided. The public interest is best served by the current process wherein a private sector entity with an independence under the law is regulated by an independent governmental agency. This interaction provides a process which extends to almost every aspect of utility operation.

(c) Source of Initiative

Another aspect of the comparison arises out of the contrasting roles, powers, and authority of a corporate executive and a manager of a public sector agency. The difference is based upon a historical and functional dichotomy which is similar to the dichotomy between common law and the civil law. The common law, derived from Anglo-Saxon jurisprudence, interprets the individual as a legal person with total rights and powers. These may be limited and circumscribed from time to time by statutes and decisions of the court, but the basic right of an individual, his raison d'etre, is to be himself. This means that he can act using any of his impulses, motivations, insights, concepts, or imagination. He is indeed a free spirit. His rights are all-inclusive except as they may be limited by specific law.

On the other hand, in the civil law contemplation the individual has only those rights granted to him by statute and regulation. He can only do what he is authorized and directed to do. He is only the agent of the governing mechanism. He is a "negative" person, so to speak, with only those rights specifically conferred upon him by law.

This means that the manager of a public sector electric agency may do what the statute permits him to do, and nothing more. He may think he can operate as his counterpart in the private sector agency does, but he is wrong. He is a very limited manager of a technical operation who is not entitled to be an entrepreneur, to plan for expansion, or to express views concerning legislative policy relating to his operation. He is circumscribed by the "civil" law under which he operates.

On the other hand, the officer of a corporation such as a private utility not only has the privilege, but also the duty to make his enterprise as profitable as possible. While he is substantially limited by law and regulation, he is still free to act in terms of his own insight. Thus, in his official capacity he has a citizen's broad role to speak out on policy issues which affect his enterprise. He also should seek to expand and remold his business, and to enter new fields. All of this is subject to the constraints of the interaction with his regulators, which is part of the genius of the system.

Thus the private sector person acts out of his own perspectives and self-interests, while the public agent must act in accordance with the prescriptions of law. For the latter, this involves all of the intricacies of due process, the means whereby legal concepts are translated into action and conduct. Accordingly, the officer of a public corporation acts in the public environment. He cannot do anything unless the statute which created his position says he can.

This flexibility makes the private sector more desirable and capable of more responsiveness and responsibility than can ever be generated by the public sector agency. Furthermore, the public agency, to the extent it can act or express views concerning a public issue, must not act contrary to the views of the executive branch of which it is a part. It must be an arm of the administration at any given time, and thus it is prone to be a vehicle of political views. For example, it can hardly be expected that the New York Power Authority would oppose or contradict the findings or the energy plan proclaimed by the State Energy Office, which is also a part of the administration. A policy debate among branches of the government is fundamentally as infeasible as litigation among them. How much clearer are the roles when the production of goods and services, or the business of society, is carried out by an independent, nonpolitical entity.

In any event, it is in this philosophical and practical context that the issue should be viewed, and not in terms of labels, preconceived notions, and suspicions too often attached to the private sector and attributed to the concept of profit. The issue should not be viewed in terms of the mere assertion of the rights of private enterprise and of private property, but in terms of the functions of institutions, the beneficial activities of property and enterprise. How do things work? is the question. And if this aspect is stressed, the benefits of the private sector initiatives which are subject to the impacts and restraints of regulation will provide the best way to accomplish the task for all concerned.

In this context, each utility company, as a component of the private sector, carries out its operations. Truly in terms of the results it yields, it is a corporation which provides public benefit. It renders an effective gas and electric service, engages in a vast technological operation, pays taxes, pays dividends and interest, meets payrolls, and participates as a corporate citizen in the activities of its communities.

This viewpoint warrants social, economic, and political acceptance. Indeed, beyond acceptance it warrants support and encouragement, instead of the negative and repressive measures which seem to be offered today. Public policy should support and encourage the utility company. And with such support, there can be positive benefits to all who are involved in the process, and a meaningful contribution to the maintenance of a free society can be assured.

Chapter 7

THE BASIC PROBLEM OF REGULATORY METHODOLOGY: THE DOMINANCE OF DUE PROCESS, ITS SHORTCOMINGS AND ALTERNATIVES

A famous legal maxim tells us that "justice delayed is justice denied." Tested by this standard, regulatory proceedings represent an utter failure. Yet, utility regulation is a principle of our society which interjects law into the complicated maze of industrial society. It is the front line of governmental interaction with the life of business, and most assuredly cannot be dismissed with simple condemnations.

Born in modest beginnings in the last part of the nineteenth century, regulation became, along with our antitrust laws, a principal mechanism to protect society's interest in an unfolding world. Clearly, regulation represented one of those giant steps forward in law, such as the creation of equity and the development of the equity courts. When the process of law became too sterile and limited, more flexible rules were needed to deal with new injustices. As equity courts evolved to provide these more flexible rules, so administrative agencies evolved to provide the relief which judicial and legislative processes could not provide.

Large-scale regulation in the United States was first attempted through the establishment of the Interstate Commerce Commission (ICC). Although the regulatory commission was not without precedent in the states, the ICC was essentially a new development to meet a difficult and dangerous problem. A dynamic and powerful industry was engaging in practices considered detrimental to the public welfare. The industry's tremendous impact on the national economy called for some kind of continuing surveillance, and the economic questions involved were beyond the competence of the courts.

The Commission was initially established to deal with the problem of discriminatory rates imposed by the railroads. It was subsequently empowered to review and fix interstate rates. The ICC was designed as an organization to be operated by people with a special understanding of the problems of the railroad industry. It was intended to provide a means of reviewing the economic factors involved in the railroad industry, particularly as they affected the railroads' customers.

The Commission was generally effective and very shortly gained genuine respect. Significantly, Congress increasingly followed the pattern of the regulatory commission rather than the judicial pattern of the antitrust laws. With this, a host of commissions and agencies with rulemaking authority was created and developed.

As a result, regulation, moving from a modest beginning, has become a massive undertaking. And while there is a continuing sense of its need, there is almost universal dissatisfaction with its process. The consumers think the regulatory agencies are probusiness; business, on the other hand, thinks that regulation is needlessly slow, tedious, and often inept.

As a consequence, suggestions are continually offered for improving the process. It has been suggested to: (1) increase the efficiency of the administrative process by providing better personnel and adopting better "in-house" organizational methods; (2) broaden the administrative process; and (3) convert the administrative process into a planning operation. But none of these suggestions, worthy as they may be, dig deep enough into the fundamental issues. None of them attack the basic problem of regulatory methodology.

The basic problem of regulation is its commitment to the procedures in which the parties are adversaries, and the consequent requirements of due process. It is, accordingly, the thesis of this chapter that the fundamental malaise of regulatory methodology is the misapplication of due process considerations and the insistence on adversary proceedings.

Administrative law initially dealt largely with ratemaking. Although rate questions have become difficult and abstract, and involve extremely subtle accounting considerations, the matter of determining rates is a function of a limited range of variables: rate base, expenses of operation, rate of return, and what can be excluded and included. It is a one-dimensional operation.

In dealing with these questions, administrative law remained judicial in character. A tribunal heard adversary proceedings which involved pleadings, taking of evidence, and the observance of the requirements of due process. Eventually it issued a judgment. The tribunal was intended to be a body of experts, and it had the capacity to deal with problems which could be solved by applying expert knowledge and judgment. Basic to the process was the premise that the problem could be resolved by a rule, a judgment, or a decision.

Today, however, social questions are multidimensional in that they deal with diverse matters such as land use, poverty, race, cities, and baffling technical considerations. Such questions defy clear and certain solutions. But solutions must be attempted, and they often cannot be anything but tentative working hypotheses, subject to continual reconsideration. An unalterable rule—even one made under the best and most flexible of hearing procedures—is often inappropriate.

Notwithstanding the increased complexity of the issues, regulatory agencies continue to operate under due process. Due process, as American as the Fourth of July and revered in judicial history, becomes sacred, exalted, and venerated. But veneration often covers up real meanings and denies appropriate applications. A blind observance of words may do more harm than good.

The human values which are preserved by observing due process are foremost in the American scheme of values. Yet, because of the extraordinary and emotional appeal of the ideas associated with it, due process tends to be applied almost without limitation or distinction to all sorts of legal situations.

The phrase "due process" has been used in so many cases and applications that it is impossible to define it. Functionally, there have been two tendencies in the use of this phrase by the courts. One is a broad and general tendency to use the phrase subjectively to provide a means whereby actions which have been characterized by certain expressions have been held in violation of the Constitution. These expressions include "unreasonable," "arbitrary," "capricious," "contrary to a fundamental sense of civilized justice," "shocking the conscience," and "offending the community's sense of fair play." The broad form has given due process its mystique. In the narrow form, the phrase has been interpreted to be synonymous with the "law of the land" and descriptive of the procedural requirements in a trial. This narrow form has given due process a guiding influence in determining the modes of procedures of litigated actions. In any event, as long as due process administrative proceedings are required, administrative due process will have to be observed. This includes: (1) the opportunity to be heard, (2) due notice of hearing, (3) fair conduct of hearing, (4) support in the record for the decision, (5) submission of proposed findings in a tentative report, and (6) the opportunity to file and be heard concerning exceptions to the report.

In effect, this means a trial that is derived in form from the common law courts. It means witnesses and counsel operating through direct testimony and cross-examination. It means a tense atmosphere in which everyone is protecting himself rather than relaxedly seeking the truth or, more aptly, the best solution. It also means a tedious and tortuous hearing in which lawyers ask questions about technical matters with which they are really unacquainted. They are advised by experts on the questions they should ask, but what is transmitted from the expert to the lawyer often becomes unclear. In any event, an inarticulate question is asked and an inarticulate answer is given; instead of clarifying, the exchange becomes garbled. The proceedings go on in this indiscriminate fashion. By the time the court comes to a decision, the information which was presented at the beginning of the proceeding is now outdated. The hopeless process goes on without achieving results satisfactory to anyone. All of this is done with a solemn endeavor to be sure that all the legal necessities are observed.

Notwithstanding all the good intentions and the scrupulous attention to due process, it is questionable whether due process is achieved in a meaningful way or merely given verbal expression in an administrative proceeding. It is further debatable whether due process can be achieved in some types of issues, particularly those involving aesthetics, scientific questions, and environmental concerns. And a more important question arises: Is it desirable to achieve due process in dealing with social questions of future policy?

In considering the application of due process in complex procedures which relate to the question of a modern industrial society, three complexities appear: (1) the nature of the issues, (2) the nature of the parties, and (3) time.

The Issues

Where clear-cut issues are to be found—where, for example, A sues B or the government imprisons or fines an individual for violation of its published and comprehended decree—the procedures of the common law before judgment is passed are totally appropriate. In these situations, the need for rigid proof which follows logical, deductive reasoning produces a rationality and a fairness which are elements of due process. In a trial, lines of proof are developed to determine whether the conditions of the rule of law have been satisified or violated. In the simplest case, where there are no questions of law, a trial becomes one of questions of fact. This consists of the processes of proof and controversy which are regulated and guided by principles of logic, rules of law, and the content of human knowledge. But the simple case, even in the common law court, hardly ever exists.

Questions of fact and law are often intimately related. The relation is often so close and intricate that it is not always profitable to attempt to separate them. Futhermore, the time necessary to conduct such strictly logical demonstrations is not always consistent with the importance of the question raised, and expediencies inevitably become necessary. Nonetheless, in the common law courts the rational steps of logic can be successfully pursued.

In complex technical matters, the clear-cut demarcations become difficult to discern and logical lines of proof become difficult to develop. Even here, the procedures of due process are reasonably acceptable as long as administrative process deals with relatively simple cases. Multiple-party rate cases, cases involving economic and environmental impact questions, and franchise cases in which private interests seek to use the public domain become more complex.

The endeavor to compress the matter into rituals of logic, as required by due process, produces a fairyland. For example, approximately two decades ago when the local rate case developed into a natural gas area rate case, the Federal Power Commission endeavored to determine complex economic matters which are of broad national interest and consequences. Another example of a complex issue is a generic proceeding attempting to settle questions of fact and law which are to be applied later to specific cases.

The proceedings become even more obtuse when the questions are those of environmental considerations (scientific and aesthetic), the quality of television programs (aesthetic), or the comparative costs of nuclear or fossil fuels (economic). In these situations, there are no simple, clear, logical due process conclusions. No due process forum can successfully be the judge of the aesthetic. No such forum can be the judge of the continually shifting values of the marketplace, nor of the appropriateness of a technical solution to an engineering problem.

In short, the variables and the indeterminates in the issues are so numerous, and matters of judgment and tastes so predominant, that applying due process in modern administrative proceedings, as attested by the history of the last several decades, has become tortuous, if not impossible.

The Parties

It is clear that the foundation of due process was related to personal matters and was derived from the need to afford all possible protection to individuals whose liberty or property was about to be impaired. Even the language of the Magna Carta makes it clear that the origins of due process related to the rights of individuals.

Arising out of and existing in simple, though completely fundamental, relationships, the issues were the socially uncomplicated prohibitions of criminal law directed against the conduct of individuals. The purpose of the concept was to make sure that procedures were recognized and understood, designed to produce fairness, and followed before the sovereign's determination was made. The concept did not relate to relationships among groups, for the concept of the corporation was barely beginning to be understood.

Today, however, the problem is complicated by the fact that the parties to an administrative proceeding are rarely individuals. They are usually groups appearing in some corporate or associative form who undoubtedly reflect a collection of interests. Even the accepted procedure of treating the corporation as an entity is not always realistic.

If, on the other hand, the attempt is made to apply the logical rigidities of due process to the components of a corporation—for example, to stockholders, to management, or to employees—the process becomes complex indeed.

Similarly, when we consider the intervenors in an administrative proceeding, we need to inquire about their true nature and identity if we truly wish to make due process workable. It is not enough to merely ask, What are the names of the organizations and what are their avowed purposes? We must find out who comprises the groups which are appearing and how truly they represent their alleged interests and points of view.

Equally valid are the questions: Who is not represented by those present? Who may be affected? Who is not present to speak? Do the titles of the groups adequately describe them, or are they fronts for other interests, possibly competitors in disguise? All of these questions should be asked if due process is to be adequately satisfied. But if these questions were fully pursued through due process procedures, the already slow administrative process would become painfully and inadequately slower. All parties would have a fine exercise in logical procedures. But by the time the proceedings were concluded, the factual situation with which the proceedings were dealing would be so changed that the issue would probably be moot.

If, however, the requirements of due process did not have to be satisfied, the question of interest or inquiry would not be asked. Depending upon the type of forum, all who wished to speak would be heard. The rationality of their statements would be the issue, not who they were or what position they represented.

Time

Politics and the natural sciences recognize the extreme importance of time. On the other hand, judicial science, for the most part, functions as if time were not essential. Where the question is a matter of penalties after an act is committed, the slow and methodical progress of judgment is totally appropriate. Yet even here, as the Chief Justice of the United States has indicated, expedient judgment is merciful to the accused as well as beneficial to society.

However, when the judgment relates to future events—for example, when a utility has to figure out, based on events that haven't happened, what rates to charge in the future—then time is as important as any other element in the situation. If, however, the decision of the administrative tribunal must be accorded the procedures required by due process, the matter before the commission may very easily be moot before the issue is decided.

For example, the financing programs of public utilities are subject to the approval of administrative agencies. The need for financing and the type of financing can be scrutinized prior to the date selected for the sale of the proposed securities. While the utility's demonstration of this need is beneficial to the public interest, the utility has no opportunity to avail itself of an element of due process—namely, the right of appeal—if it feels aggrieved by the commission's order. By the time the appeal is effected, the condition of the market which made the timely offering will have changed. It is a rare instance, indeed, when a commission's financing order is appealed. For the most part, the utility accepts the order with whatever unpalatable conditions it may contain or withdraws the application. Because of time, however, the utility is effectively denied the due process protections represented by an appeal.

Likewise, if the administrative agency seeks to deal with a matter of engineering or scientific endeavor—such as a utility wanting to build a demonstration coal plant using scrubbers—it is invariably dealing with situations where time is important. Licensing procedures before the Nuclear Regulatory Commission, applications before state siting boards, and proceedings relating to safety questions all move at a snail's pace. This leisurely pace seems to imply that the construction of the project involved has no time considerations. Truthfully, however, nowhere is the adage that "time is money" more applicable.

Notwithstanding the importance of time, due process requires the regulatory agencies to follow procedures which, under Federal Energy Regulatory Commission rules, for example, involve these steps:

- Application
- Public Notice
- Petition to Intervene
- Answer to Petition to Intervene
- Order Fixing Hearing Date
- Pre-hearing Conference
- Hearing Briefs to the Presiding Examiner
- Reply Briefs to the Presiding Examiner
- Initial Decision by the Presiding Examiner
- Exceptions to the Initial Decision
- Reply to Exceptions
- Commission's Opinion and Order
- Petition for Rehearing
- Final Order

Thus, the mechanics of the debate seem to become more important than the substance, and certainly become more important than the question of time which, even theoretically, is not considered an element of the procedure.

A Solution: Consultative Law, an Alternative to Due Process

If the government, as the representative of the public interest, is to have any role in matters where issues are complex, parties are numerous, diverse, and possibly unrepresentative, and where time is of the essence, procedures must be devised which can offer a means of handling complex issues and numerous parties in a timely fashion. In such proceedings, the emphasis should be upon objectivity and establishing an atmosphere of good faith. To establish such goals, the adversarial aspects of proceedings should be minimized and the investigatory characteristics emphasized.

In short, a method of avoiding due process should be devised. On the surface, such a suggestion seems to be a startling conclusion, inconsistent with our history and our cherished ideals. To abandon due process is to mock the past and almost literally to have the blood of the martyrs on our hands. No one would want to turn back the pages of history to undo what has been so laboriously developed at such a great price. Obviously, nothing so drastic is being suggested, but distinctions are in order. All legal proceedings do not have to require the functioning of due process. Obviously, due process must be satisfied in any situations where an agency issues an order which compels the taking of or refraining from action. But perhaps there are ways for government to proceed which will avoid creating situations where the issuing of an order and the procedures of due process are necessary.

And indeed, such procedures began to evolve during the late 1950s and the 1960s because nothing else could be done. The key to the procedure is the "sidestepping" of due process. The government is involved and is part of the process, but it does not produce an order; accordingly, the formalities of due process are not required.

The process suggests an advisory procedure or consultative law. "Consultative law" is a term developed by the author. It means developing procedures whereby commissions and other governmental groups can enter into consultation to produce settlements. Although not formally recognized, the process takes place regularly where settlement procedures are involved. The most striking illustration is the settlement of the natural gas area rate cases two decades ago by the Federal Power Commission.

Some Examples of Internal Due Process—The Progress of the 1960s

As a result of the Supreme Court decision in the *Phillips* case, which made natural gas producers subject to the Natural Gas Act, the Federal Power Commission (FPC) was inundated with certificate and rate applications. The FPC, through formal adversary proceedings, was unable to keep pace with the applications which were presented to it. For example, in February 1961 the backlog of pending and unresolved gas rate cases before the Commission amounted to about 4,000. At the same time, there were 193 pending and unresolved gas certificate cases relating to 5,761 miles of pipeline, representing a total estimated construction cost of $850 million. Thereafter, the Commission began a course of settlements. In its 1962 Annual Report, the Commission described substantial progress in reducing, through settlement procedures, the backlog of existing rate cases. These settlement procedures were not considered to be extraordinary or an unusual process. But, in effect, they were a substitute for pursuing due process requirements to their ultimate conclusion.

The program of the Federal Power Commission during the 1960s for an advisory relationship with the electric utility industry is another excellent example of a cooperative procedure involving the government. Under that program, the FPC established a National Executive Advisory Committee and 12 regional and technical advisory committees. As a result of its work with these committees, the FPC produced its 1964 "National Power Survey," which was an attempt to survey the national electric power resources and to indicate the potential for nationwide development of those resources.

In connection with the survey, the Commission made it clear that in securing the cooperation of the industry, it would produce guidelines to foster the industry's development and would not endeavor to mandate such guidelines. It assured the industry that the Commission did not intend to preempt the basic responsibilities of management in order to plan and develop national utility systems.

The FPC appointed a new Executive Advisory Committee late in 1965, and six regional advisory committees in 1966. In creating these committees, the Commission reemphasized that they would assist the Commission and not preempt management responsibilities. These new committees were created to update the 1964 survey, develop new guidelines made necessary by changing technology and circumstances, and to extend industry projections to the year 1990.

Subsequently, after the 1965 Northeast electric blackout, a massive national reliability act was proposed. After much debate, the difficulties of dealing with the technical questions of reliability through governmental procedures became evident and the act was not adopted. Instead, the utilities established a National Regulatory Council and regional regulatory councils in which governmental agencies play a significant role. The system of reliability councils has worked very well. Notwithstanding the huge growth in electric systems, especially prior to the 1973 oil embargo, blackouts have been rare and sporadic. Even the major Con Edison blackout of 1977 remained localized and did not affect the interconnected systems.

An outstanding and clear-cut use of consultative law, explicitly provided for in statute, was attempted in the Hudson River Valley Commission law in New York State. The statute creating the Commission was passed in 1966. The Commission's purpose was to encourage the preservation and development of the scenic, historic, recreational, and natural resources of the Hudson River Valley, and to encourage the full development of its commercial, industrial, agricultural, and residential resources. The statute directed the Commission to coordinate the best plans of private organizations and all levels of government. The Commission was to assist public and private agencies in undertaking projects in accordance with a coordinative, comprehensive plan. It was intended that the plan would include recommendations for public or private projects of major benefit to the area, and for appropriate governmental units to establish zoning and other environmental control measures.

To accomplish these purposes, the statute provided procedures for preliminary consultation with the Commission by anyone proposing to undertake a project which fell under the meaning of the statute. Any project within one mile of the shore of the Hudson River, and also some projects visible from the river and within two miles of the shore, had to be submitted to the Commission for review. On preliminary review, if the Commission found that the project substantially conformed with a coordinative, comprehensive plan that was generally outlined by the

Commission, there was no further review. If the Commission found that the project did not substantially conform with the plan or would have an unreasonably adverse effect upon the resources referred to in the statute, the Commission could issue an order delaying the beginning of the project and scheduling a public hearing within 30 days. On or before the hearing, the Commission reported its findings to the agency or person proposing to undertake the project, and to any public agency having the power to review or approve the project. The Commission was instructed to see that its findings were widely disseminated to the public.

Thus, the Commission could not compel or prohibit any specified action, but it could cause delays for a short period of time in order to permit the marshalling of information and competing viewpoints. Making this information public was expected to focus the pressure of public opinion on a proposed project, while at the same time providing recommendations and advice to any authoritative governmental agency, such as a zoning board.

Although the Commission's life proved to be very short, there were several cases where it was effective in obtaining voluntary compliance with its views—most prominently, in abandoning a nuclear project after the Commission's findings regarding the 766-megawatt nuclear power generating facility of Niagara Mohawk Power Corporation. The Commission offered a promising process for obtaining a consensus which would accommodate technological, economic, environmental, cultural, and social views. During its existence, the Commission operated with considerable success. But then it was absorbed into a statewide agency, which was created when environmental interest became more acute.

Another example of the legislative adoption in the 1960s of the consultative procedures was the Air Quality Act of 1967. This act was not purely consultative since it contained provisions for injunctive relief and penalties. However, a substantial portion of the law provides for the consultative process, and the injunctive relief was authorized only where the emission of contaminants resulted in an imminent and substantial danger to health. In its mixture of traditional remedies and newer procedures, the law demonstrated the flexibilities which consultation may provide.

The Retrogression in the 1970s

With the close of the 1960s, the momentum to move in simple commonsense directions seemed to be halted, and there came a return to the due process commitment in most inept and difficult situations. The balancing of considerations which relate to environmental questions clearly requires a simplified consultative approach, but instead both the Federal Clean Air Act and the Clean Water Act are procedurally regressive.

The Clean Air Act was adopted basically in its present form in 1980. Its primary purpose is to protect and enhance the quality of the nation's air resources in order to promote the public health and welfare and the productive capacity of the population. The Clean Water Act was adopted basically in its present form in 1972. Its primary purpose is to restore and maintain the chemical, physical, and biological integrity of the nation's waters. Goals of the Act are (1) to achieve water quality which provides for the protection and propagation of fish, shellfish, and wildlife, and provides for recreation in and on the waters and (2) to eliminate, by 1985, the discharge of pollutants into navigable waters.

Both of these massive acts are wordy, regulatory in language, and quite unartful as statutes. Both provide for due process hearings at an administrative level followed by the possibility of court appeals.

Throughout, the Acts reveal the governmental career bureaucrats' conviction that American capitalistic business is predatory, especially in its disregard and destruction of the environment. All of this is said without recognizing that industrial processes in the United States were conducted in accordance with the understanding and mores of their times. Only the apparent affluence of the late 1960s led to the possibility that the economy could afford the environmental cleanup. Thus, the Acts are punitive rather than constructive, adversary rather than consultative. Once again, scientific issues are subjected to inept legal due process scrutiny.

The results have been horrendous, leading to almost complete dissatisfaction. Rarely has a sense of balance been displayed, and the question of cost/benefit has been discarded. Instead, a bureaucratic determination to produce a zero-risk environment persists, and the process of producing this result by governmental order continues.

In New York State, the so-called Article VIII proceedings represent the same approach. Article VIII is an additional article to the Public Service Law in New York, and provides a siting procedure for electric power plants in this state. No plant subject to the act may be built without obtaining a certificate of public need and environmental capability issued by a siting board created under the statute.

Though the act has recently been amended, the original act led to long, complicated, and costly due process proceedings. Throughout the five years of the existence of the unamended act, no certificates for power plant construction were issued. Luckily, the demand for power in New York State rapidly diminished during the years in which the original act was in existence so that no crisis was encountered. Still, the act is a monument of how not to proceed when considering social, economic, and environmental questions that relate to future events.

Consultative Law: The Hope of the 1980s, The New Approach in Public Interaction

Quite clearly, a new approach is needed. Precisely, it is indicated in situations where: (1) the complexity of the problem exceeds the capabilities of existing procedures; (2) the desirability or possibility of formulating a uniform rule is unclear; and (3) the entities involved have a sense of a broad community responsibility.

The precise form of consultative proceedings is unimportant; the theory and approach are significant. The basic issue arises out of the fact that in our political system, the government cannot issue orders willy-nilly. Yet this simple fact is strangely misunderstood. A common misconception appears to persist that a wise government "knows" and "tells"; the truth is that government neither knows nor has the power to tell or to issue orders. It can do this only after due process has been afforded, which usually means after an adversary litigation. The trick, then, is to avoid the confrontation and to seek common understandings and agreements. Consultative law seeks to avoid the "order" and the controversy.

In consultative proceedings, because the theory and approach are significant, methods and procedures can be devised and adapted as situations require. The methods that appear possible are:

1. Settlement techniques
2. Consultative forums
3. Representational proceedings
4. Associational proceedings.

Settlement Techniques

Settlement techniques have been used in several situations. Settlement procedures have basically been "ad hoc," and have been developed as emergencies have required. Specific statutory procedures providing for settlement could be devised.

For example, in a utility rate case, the statute could provide that all parties to a proceeding be required to meet in a conference-type meeting. No witnesses would be sworn in to submit their testimony in artificial question and answer form, and then subjected to tedious, irrelevant, nonproductive cross-examination. Instead, one or more representatives of the company seeking the rate increase would make statements in considerable detail, supported by exhibits consisting of tables, charts, and graphs. Informal clarifying questions by all present could be asked and the data presented would be available for review by all concerned over a two-week or one-month period.

During this period, the governmental agencies and intervenors involved would intensely scrutinize the material. At further meetings they would indicate, either by written declaration or by oral statements, what areas they wished to dispute and what areas appeared acceptable. If everything were in dispute, then the proceedings would have to move on in the traditional way to reach a judgment. But if great areas were indisputable, which is very often the case, then the areas in dispute could be assessed, the dollar values determined, and simple monetary compromise could be negotiated against the framework of the entire proceedings.

Principles involved in specific issues would eventually be decided on the basis of the philosophic outlook of the adjudicator, tempered by the precedents within which he was working. Readily determinable dollar results would be produced, to which the parties would agree without an order. Then, failing agreement, adjudicatory proceedings could be pursued.

As an alternative to the procedure just outlined, the governmental agency involved, after it had heard the company's case, could develop a document of settlement which would then be submitted to all parties for review and comment. If the document, with or without adjustments proposed by the parties, were acceptable, it could become the agreement of settlement; if not, the matter would move on to a judgment.

Consultative Forums

To provide for a governmental forum, a governmental body (or private bodies, groups, etc.) would conduct a forum at which the judgments of responsible and interested parties could be tested. It would have no compulsive power, except to require the submission of information and data pertinent to the issue being explored. It thus becomes the agency which insists that a total community reaction be brought to focus upon the problem. Furthermore, this forum is not one in which partisan claims are judged, but one in which viewpoints are submitted in good faith to develop the best possible solution through consultation.

Representational Proceedings

In another possible forum, governmental committees would marshal the considerations pertaining to the general public while industry committees would present the industry view. Competent staffs would be provided for each committee so that appropriate reports could be presented and required research undertaken. Parties which may have an interest could also be heard. It may be possible that they would not be directly heard, but would have their view represented by either the industry or government committees involved. The right of personal appearance now accorded in administrative procedures must give way to the practical need for expeditious resolution of differences in opinion. In those areas where an "order" is neither required nor possible, the administrative proceeding should give way to an interaction between governmental committees that express the public view and industry committees that express the enterprise point of view.

Indeed, if the issues could be adequately stated, the committees representing special interests such as environmental groups, government, or companies, could engage in a form of collective bargaining. The novel structure of our labor laws which prescribe no result and provide for no substantive orders offer a form of consultative law. The only requirement of labor laws is that the parties bargain in good faith. This platitudinous criteria, which cannot call forth any enforcement, has worked surprisingly well. Bargaining is done in good faith, and bad faith becomes apparent quickly.

In similar fashion, the representational committees could bargain and could develop agreements. If they were not forthcoming, then traditional judicial findings could be employed. If any such system were in effect for any period of time, procedures would be developed and traditions evolved so that the bargained result could easily become the norm rather than the exception.

Associational Proceedings

It is not essential that the consultative process operate under a governmental aegis. Located in New York State's Mid-Hudson River Valley, the Mid-Hudson Pattern is a nonprofit corporation supported by contributions from its members and the general public, and by proceeds from various contract services which it can supply. The purpose of the organization is to investigate and analyze issues, and to develop facts relating to the impact of proposed physical facilities (such as a factory) upon the social and physical environment. Questions asked include: Will it pollute the air or water? What will be the impact on the school system and on roads?

In a sense, the Mid-Hudson Pattern is a planning agency without a portfolio, without an entity to plan for. But it brings the facts relating to any given proposal before the people of the communities it serves: the corporate entities, the businesses, the governmental agencies, and county and town boards. The organization gathers information, prepares professional staff analyses, conducts public forums, and provides the information and exchange of viewpoint so intelligent lines of action can be undertaken by public and private sector agencies.

Local municipal governments and private agencies regularly consult with the Mid-Hudson Pattern staff and bring its processes into operation. The organization has been modestly successful, and undoubtedly could be far more so if adequate staff and funding were available. In a sense, it is ahead of its time. The concept offers a further potential to unfold the areawide decisional process without resorting to law or tribunals and courts.

Conclusion

Many types of proceedings could be developed and the methods of consultative law would be flexible and adaptable as circumstances required. Whatever its form, consultative law has one single characteristic: to express a liberalizing of due process requirements. Perhaps, indeed, it should be called non-due process law.

It will also find government in a different role than its traditional one. For in modern society, a new and larger community of interest has developed. In effect, a sociological group has been established, and this group can be looked upon as the community itself. On one hand, the parts of this community are private interests which, because of their significance, become public in their impact. On the other hand, they are public interests which are expressed by the organs of government. These groups interact with each other and produce action and motivations which require decisions and resolutions.

A smaller prototype of this larger community is the business corporation in which diverse interests oriented to a single objective produce decisions through a process of consultation and conference work. It would be unheard of for a corporation to decide its policy by some kind of internal court. And yet, with all its faults, the corporation has been the effective instrument for developing our economic resources. Part of this effectiveness has been due to its ability to move relatively quickly and with flexible procedures.

In a similar fashion, consultative law has the potential of producing a process internal to the community in which the considerations of the community can be expressed in a sensible, practical way. Consultative law will be free of the burden of developing the logical symmetries of due process requirements. It will recognize that all interested parties will not be physically or personally represented. This will be more realistic than the present administrative procedure which pretends to be logical and inclusive, yet finds itself incapable of being either. Instead, it will strive to find those areas where government order can be dispensed with. In these areas, it will try to cultivate a discourse which will produce a rationality rather than the advancement of interests. Where an agreed-upon rationality cannot be produced, then a flexible accommodation, subject to change as situations and times unfold, will be developed. Nor will consultative law seek the "right" solutions. In the shifting character of matters affecting aesthetics and the environment, for example, there are no right solutions. There is only the best solution in view of all the circumstances. It will foster a process of an endless search for a rationality, objectivity, and accommodation of viewpoints.

In these circumstances where the milieu of social conduct is so uncertain, there can be no segment of the community, such as the government, which can draw itself above the rest and say, "This is the rule, this is the order." No process of known procedures can produce this result.

On the contrary, however, if the government can look upon itself as an element of the community, it can use the forces which it can command to produce a new process in the community.

While the way is far from clear and only the basic outlines are beginning to emerge, consultative law offers a possibility of new way of community action. Potentially it may open up, in the democratic tradition, a way in which government—the one community entity which ultimately derives its sanction from the people expressed through the elective process—can become an organism of leadership and persuasion rather than compulsion.

The development of the techniques necessary to bring this about is slow and the means of adaptation tedious. But there is much to suggest that legal energies should not be directed so much toward the perfecting of the means of due process necessary to produce an order as they should toward perfecting the techniques of producing the result without the order.

As society evolves and develops, the role of the order and the injunction should be less pervasive. The operation of the public utility is in a sense a laboratory for this process. No other institution has the sophisticated experience of the public responding to the demands of the public interest. Clearly, however, it cannot continue to struggle with current tortuous methods of regulation. A better way must be found. And if it is found, then all the experience of the utility industry and the regulators can be brought to bear upon the problem of the appropriate development of economic activity with due regard for the public interest.

If the social interactions which the utility industry represents can be made to operate in an atmosphere of good faith, the utility industry may become a prototype for future American development.

Chapter 8

COMPETITIVE QUALITIES OF THE UTILITY PROCESS

A unique quality of a human being is the ability to speak, write, and read. Words are the mechanisms of human thought. When the Constitution guaranteed the right of free speech, the Founding Fathers were expressing an essential component of political freedom that was not only part of their inheritance, but a reaction to the real or imagined restraints imposed by the British Government.

The Founding Fathers guarantee of free speech was also an article of faith. Beyond these two considerations, free speech means the essentiality of the human spirit; to be restrained in what one can think or say denies the human condition.

Thus, the concept of free speech is extended to corporate life to the individuals comprising it. The possession of the right of free speech by utility companies recently was confirmed by the United States Supreme Court in two cases: *Central Hudson Gas & Electric Corporation* v. *Public Service Commission,* and *Consolidated Edison Company of New York, Inc.* v. *Public Service Commission.*

In these cases, both decided on June 20, 1980, the plaintiff utility company was subject to an order of the New York Public Service Commission. In the *Central Hudson* case, the order prohibited promotional advertising; in the *Con Edison* case, it prohibited bill enclosures which contained political messages. (A summary of the *Central Hudson* case is included at the end of this chapter.)

Several issues were involved in both cases, but one of the central questions raised was: as a monopoly, is a utility entitled to the free speech protection? This profound and provocative question raises the cardinal issue of free speech and also begs the question of monopoly. It reveals the general assumption that a utility, though a monopoly in some aspects, is monopolistic in all.

Electric utilities enjoy exclusive franchises to provide electricity in their franchise areas, as do gas and water companies. Within the perimeter of its particular service, a utility is a monopoly and has exclusive rights. In exchange for these exclusive rights, it accepts regulation and limitations on earnings.

From a product standpoint, a utility company is a monopoly; otherwise, it is competitive just like any other business. In the area of values and ideas, for example, it is challenged more competitively than other businesses.

In deciding that the utilities were indeed entitled to free speech protection, the Supreme Court confirmed the utilities' rights of expression, speech, and communicating—in essence, the right to be competitive.

Utilities are competitive in many aspects, and have the potential of becoming vigorously competitive. There are five ways a utility is competitive: functionally, in value, institutionally, technologically, and in ideas.

Functionally

An electric utility, for example, actively competes with gas, oil, coal, wood, and solar companies in the space heating market. In air conditioning and in some appliances, such as clothes dryers, electric and gas companies compete. In industrial and commercial processes, electric utilities compete with gas, oil, and coal companies. Electric utilities are also competitive in various forms of cogeneration. In transportation, the electric locomotive competes with the diesel, and at one time it competed with the steam engine. Also in the past, the electric streetcar competed with the railroad. In the future, the electric car will compete with the gasoline and diesel automobile. Thus, although a utility is the exclusive supplier, its product must compete with others which can accomplish the same objective. And if the utility is viewed as a provider of energy services such as comfort, light, mobility, etc., it is clearly a competitive institution.

In Value

Utility rates are established to provide an appropriate return on invested capital. The effect of utility rates can be measured against cost-of-living indices. During the past several years, utility rate increases have generally been slightly less than the increases in the cost-of-living index. If the fuel adjustment component is eliminated, the increases in electric rates have been substantially less than the cost-of-living index.

Thus, utility rates can be measured against the charges for other goods and services in our economy. The objective of a utility company should be to keep rate increases in line with the rises in the cost-of-living index.

The competitive impact is also seen in the price elasticity which utility companies have experienced in the last several years. As volume sales have decreased with increasing prices, customers have chosen other products or other purposes than utility use. They have used their dollars for other objectives than electric service.

Electricity provides an absolute value which cannot be measured or provided by any other method. If the work performed by electricity is compared with the same work performed by other means, particularly manual labor, the value of electricity is demonstrated enormously. Electricity has an immense competitive advantage over innumerable alternatives. Indeed, it is universally consumed because it is so much more useful and economical than its competitors.

Viewed in this way, electricity is sold at an artificially low price. Perhaps consideration should be given to allowing its price, at least in some of its uses, to rise competitively to the level of its alternates.

Institutionally

The private sector utility has always lived with the threat of replacement by or competition from public power. See Chapter 6.

Technologically

Electricity is unique in its technological characteristics because it is not storable. Because electricity is constantly on demand and is supplied instantly, it requires a 24-hour-a-day operation. Like any other business, the utility company faces technological innovations, and it must remain competitively *au courant.*

This competition, and the desire to achieve the lowest possible price, has carried the industry through several phases of generation types. Coal generation in relatively small units was the starting place, but technology has opened the way for hydroelectric power, oil, gas, and coal to be used as fuel. The economies of scale have produced generating units of greatly increased size. And the immense technology of nuclear science has produced the nuclear plant.

Now the various soft technologies offer the utility company further competition. They promote nonelectric station generation, neighborhood association, and individual generation—all ways of using electricity more efficiently so utilities will not have to build any more power plants. Rather than have a large utility serve an area, there could be neighborhood associations owning their own fuel cell, with 20 blocks or so being self-sufficient.

The utilities must explore the potentialities of these concepts and, perhaps, even be involved in their development and exploitation. Their mere existence provides a competitive threat to which the utilities must respond by getting greater efficiencies out of their own processes.

In Ideas

The whole country is involved in a fermentation of ideas. Perhaps the challenges of the current energy situation will stimulate new and different responses on how to solve the energy issues confronting us. But the clash of ideas does not only deal with the technological and economic means of response. The country is not only concerned with how to develop alternate fuels, lessen dependence on Middle Eastern oil, or manage the economic dislocations caused by inflationary energy costs. The turmoil becomes more ideological and deals with areas of values.

Our social structure is called into question and the goals of our culture are reexamined. One has to ask: Is my best interest served by an increasingly technological environment and a more specialized and disciplined social structure? Is cheap energy a sound goal? Or should we be seeking a simpler, perhaps more rural, climate than the urban structure which is the apex of our civilization?

These fundamental questions are being raised, and the utility company receives the impact of the unrest which these questions produce. Anti-nuclear demonstrations, with all their nonnuclear overtones, are directed against utility companies. Utility companies must respond by offering a steady, intelligent, effective response. This reaction, regardless of the ultimate outcome, will be a worthy foil to the institutional system which the utility represents.

Thus, while the competition of ideas intensifies in debates, arguments, writings, publications, and political controversies, the utility company sees that competition in terms of overt acts and programs. Demonstrations, media programs, and opposition in rate cases are expressions that seek to thwart the objectives the utility believes are necessary to appropriately serve the public. The utility company must respond with the rationalization of its position, both in terms of limited short-range goals and long-range objectives.

Summary

Although a utility is regarded as a monopoly—and indeed in certain aspects it is a monopoly—it is functionally an entity subject to many competitive forces. As a private sector enterprise, its genius is that it retains the competitive need to survive. Its existence is by no means assured or guaranteed. The utility company is a self-contained budgetary unit which must, within the bare corners of its own financial statement, show its fiscal viability.

A utility has only one source of revenue—its customers. It has no access to subsidies or tax revenues, no way of sharing any of its deficits over a broader social range. Thus, it must price its service at rates which are commensurate with its total costs of service. Perhaps it should have far greater flexibility in determining its own rates.

Notwithstanding the sophistication of direct regulation of rates, it is still a very tentative and judgmental process. What is just and reasonable is difficult to determine apart from the marketplace. Instead of sterily attempting to determine appropriate rates of return, perhaps regulation should concern itself more with competitive values than the logical deductions implicit in return on rate base determinations.

One way to accomplish this purpose is for a utility company to supply testimony in a rate case which compares the rate increase requested with increases in the cost of living index. If the rate increase is in line with the increase in the cost of living increase, the rate increase is competitive. If it is below the increase in the cost of living index, which, especially when the fuel component is eliminated, has been the usual experience, the utility is outperforming the economy.

A comparison of utility charges with the charges of all other products provides a judgment by competitive values. Such awareness would tend to produce a more favorable public perception and avoid the negative impression implied in monopoly.

As an example, rate forms could become more flexible. A company with overcapacity could be permitted to provide mild discounts to stimulate—temporarily—greater use and lower unit costs. The changes in rates between blocks could vary from time to time to meet varying use patterns.

And while residential rates will undoubtedly have to be regulated, perhaps industrial, and in some cases commercial, rates could be deregulated and permitted to rise to the level of costs of self-generation. This would require cases to allocate appropriate costs to the jurisdictional and nonjurisdictional business, but such proceedings are hardly novel or insurmountable.

The aim of the companies and regulators acting for a common cause should not be to determine how little the utilities need to get by. Instead, they should ascertain the proper charge in order to make the utilities fertile and ebullient.

It is not difficult to demonstrate that the railroads, subject to a regulation of paucity, were not able to get the capital required to make railroading responsive to increasing needs and competitive forces. Had the railroads been able to keep rolling stock, roadbed, and other facilities in top-level shape, many of the energy and transportation problems of today would be substantially reduced.

Therefore, a breath of fresh air of competitive thinking is needed to invigorate both the utility companies and the regulators. Utility companies must strive to gain the greatest range of freedom of action; they must have the freedom to explore and express ideas and then they must generate the vigor and initiative necessary to exercise their freedom.

At the same time, legislators, regulators, and the public must recognize the need that utility companies have for an environment where utilities can express their professionalism and desires to enlarge the public interest.

It is the challenge of the private sector to reach out for the better, the improved, and the superior. And under the competitive challenges just described, the utility must reach out for this excellence to meet the challenges—not only to survive, but to continue to render to society the benefits which have been discussed in Chapter 5.

Summary of the Supreme Court ruling in the *Central Hudson Gas & Electric Corporation* v. *Public Service Commission* case.

In December 1973, the New York Public Service Commission ordered electric utilities in New York State to cease all advertising that promoted the use of electricity.

Central Hudson Gas & Electric Corporation argued that the Commission had restrained commercial speech and violated the First and Fourteenth Amendments. The Commission's order was upheld by the trial court, the intermediate appellate court, and the New York Court of Appeals.

The Supreme Court, however, reversed the judgment of the Court of Appeals and held that the order of the Commission violated the First and Fourteenth Amendments. The Court held that Central Hudson's monopoly position does not alter the First Amendment's protection for commercial speech, and that in the absence of a showing that more limited speech regulation would be ineffective, suppression of Central Hudson's advertising cannot be approved.

Chapter 9

THE UNWRITTEN AMERICAN CONSTITUTION

In this country, when we speak of constitutional questions, we are usually referring to issues which arise out of our written and defined Constitution. Yet in the British sense of the unwritten constitution, we, too, have a basic unwritten one. That unwritten constitution both undergirds and flows from the written Constitution.

While our written Constitution does not speak of free enterprise, free enterprise is inherent in any system in which all powers not reserved to the federal government reside in the people, and in which the people have the right to carry on enterprise both individually and collectively. The constitutional reservation also implies the right to institute and carry on free economic, social, and cultural activities independent of the state. Thus, the American scene is characterized not only by business, but also by such independent economic enterprises as private schools, churches, trade unions, and artistic societies.

Our unwritten constitution provides for the public sector and the private sector. Originally, the private sector was by far the more important. Only a few people were engaged in government and only a small part of the dollar flow involved the government. The process provided for a fantastically rich and thriving country. At the same time, the dominating private sector fostered a great number of abuses. The stream of inhuman conditions which resulted from pure laissez faire is all too well known. And periodically, the system seemed to break down and needed governmental support to mend it.

The biggest crisis was the Great Depression of the 1930s. Whether the private sector could have straightened itself out without The New Deal is debatable; but that the private sector system did not avoid the debacle is hardly a challengeable fact.

A social revolution has occurred since 1930 in the whole fabric of our society. While the consequences of that social revolution leave considerable doubt about its value, the fact that the revolution occurred cannot be denied.

In the decades following the Great Depression, the United States experienced the New Deal, World War II, Kennedy's call for New Frontiers, and Johnson's Great Society. Out of these experiences there has been a vast proliferation of bureaucracies and governmental interventions and regulations, especially during President Carter's tenure in office.

All of this tends to a simplistic concept that government is always wrong. But you can't blame the government for the depression and for many other economic phenomena; the private sector has had a great share of responsibility for these problems.

It is nostalgic to think of the 1880s, with all the freedoms and lack of governmental restraints which characterized that time, as marvelous, halcyon days. But it does no good to talk about going back to the "good old days."

Today, government and business are struggling to determine just what environmental safeguards should be required. In the view of many, imposing these safeguards is considered an unacceptable intrusion of government, but other forms of governmental restraints—safety and fire rules, for example—were resisted in the distant past, but are now considered to be normal and expected.

It is difficult to think of fire laws being resisted, but at one time they were as novel as air and water controls are now: Even in the early days of fire laws, cost/benefit and technical problems were part of the picture.

Until standards for fire codes and regulations had been in effect for a period of time and the relationships among all parties involved had been settled, there were many objections to compliance with the fire regulations. From time to time, exceptions and relaxation of standards had to ensue. In New York City, for example, a statute was passed in 1761 which ordered that after January 1, 1766, all buildings in the city of New York had to have roofs of tile and slate. Wood and thatched roofs were outlawed. But, owing to shortages of tile and slate and a pressing need for housing and shelter, the law was suspended twice. Nonetheless, it finally was enforced and came to be regarded as necessary and essential.

In fact, it is quite impossible in a competitive system for safety and environmental costs to be included in the cost of a product without some governmental or cooperative restraints. If, for example, a socially-minded manufacturer includes pollution control equipment and his competition does not, the socially-minded manufacturer will probably go bankrupt. In a competitive market, social costs will, of necessity, be kept at a minimum.

Thus, every business is touched with a public interest. The land, water, and air cannot be utilized without concern about impact upon other interests. At the very least, imperfect aspects of economic transactions and the maintenance of markets require the imposition of community standards and a method of enforcement.

Government agencies often are frustrating; they often operate unwisely, and they often tend to promote the vices of bureaucracy. But, governmental agencies represent community concerns which tend to be neglected. Thus, the Securities and Exchange Commission, the Environmental Protection Agency, the Federal Trade Commission, the Department of Labor, the Occupational Safety and Health Administration, and others, together with their state counterparts, bring the impact of governmental standards upon entities operating in the free market.

In this context, all businesses—not just regulated industries—are interwoven with the public interest and must be prepared to deal with the process in which regulated industries have been involved since their enfranchisement.

This process must not be criticized or evaluated in the bland assertions of free enterprise, the rights of property, etc. Instead it must be viewed in terms of interactions and interchanges, the traditional flow of processes in which the goals of the independent free market and budgetarily-separated corporate units are tested, measured, and restrained by community requirements.

In this process, utility companies have had a long history of experience. The interaction of the utility company with the governmental mechanism can be a prototype for the development of the American system in the future.

The functioning of a utility executive as he threads his way through the many considerations which involve his company illustrate the process. This executive, not commonly thought of as an expert in political and social processes, nonetheless is in the stream of all the social considerations which exist in modern life.

In the first place, he must provide an extremely complicated technical plant, which not only generates electricity but also provides for its transmission and distribution. As noted in Chapter 5, the generating plant has evolved through periods of change and experimentation. Fossil fuel plants burning coal, oil, or gas have become monumental examples of complexity. They are dwarfed in sophistication only by nuclear plants, the products of the esoteric disciplines of mathematics and physics. No executive, of course, can be thoroughly knowledgeable about these processes, but he must have a basic grasp of their functions. And he must be able to select and work with those experts who can design and operate those plants.

Distribution systems are relatively simple in a technical sense, but transmission systems, as they grow in voltages, begin to encounter complex questions about engineering design, safety, health, and noise. A recent decision by the New York State Public Service Commission concerning the 765-kilovolt transmission lines of the New York State Power Authority supplies a measure of the complexities involved. It is a compendium of the issues, many of them controversial, which surround and relate to the construction and operation of transmission lines.

The technical matters—complex as they are—are only the foundation of the corporate structure. Out of these technical marvels, community, regulatory, and legal issues immediately emerge.

Plant siting also raises a host of technical and social issues. Where will the generating plant be located and what will be its impact? In the first place, plant siting requires choosing a location in relation to the areas where the load will be used. Proximity to the load center and to transmission facilities is the first concern. Facilities (water, rail, and highway) to transport fuel and disposed material are primary considerations.

One can no longer think that the land, air, and water can be used or disposed of as adjuncts to the rights of use which come with the ownership of property. Land use involves questions of zoning, which can either be handled locally through zoning boards, statewide where state siting agencies are involved, or federally where federal licenses are required. Air use must conform to local, state, and the federal Clean Air Act standards and the regulations of the Environmental Protection Agency. Water use is regulated in a similar fashion.

The location of the plant has an impact upon tax questions involving local communities. The kind and degree of taxation, both in the present and the foreseeable future, must be considered.

A sociological analysis of the area in which the plant is proposed must also be considered. What is the nature of the surrounding communities: Urban? Rural? Industrial? Agricultural? What kind of road system is in existence? What are the school, community, and health facilities? All these questions must be analyzed and the impact of the new plant forecasted.

On the surface, these issues—land, water, air, and community use—seem to be technological and subject to description and analysis. But underlying the technical level, there exists a value system which must be deciphered. What are the attitudes of the people in the area? Do they want growth and change? Or do they wish to preserve the area just as it is? If they wish to preserve the area, can the plant be sited without any impact on the environmental and sociological characteristics of the area? Are there differences of opinion on all these issues in the community? If there are differences, what viewpoint will ultimately prevail?

In addition, are there political impacts on the area? Are there competing industrial considerations? Is the level of employment high, or will new jobs be welcomed?

The utility executive must consider these questions and more, including many that haven't come to the surface. And in reviewing these questions, he finds himself and his company at the forefront of those issues which will shape the community. And all of these issues pertain to the development of the generating plant.

Equally important is the distribution of the plant's output. The distribution system involves considerations of franchises and the right to have lines and poles within the boundaries of the streets and over private property. Distribution is concerned with rights of way, street permits, and questions of condemnation and eminent domain.

When the plant is functional, the economic questions of costs and sales must be considered. This involves the state and federal regulatory commissions, rate cases, and all the vast expert paraphernalia which is involved in them. Although rate cases are based on the premise that a utility service must be economically viable, the cost of the system must come from the users. This concept of self-sufficiency is the cardinal principle of the private sector system and one of its key social justifications.

Does it make good social sense to have the system be self-sufficient, or should it be subsidized? Obviously, proponents of the private sector system think that it should be self-sufficient. Still, the issue is one which the utility executive must grapple with in rationalizing his business's existence.

Thus, the rate case brings up the question about how the costs should be raised and how they should be distributed among customers. Should the basis be strictly cost? If so, should the costs be historical or futuristic (marginal)? Should there be social considerations favoring the elderly, the poor, the small user?

Each of these issues has economic, commercial, and political overtures which must be discerned, analyzed, and factored into the total corporate response.

These overtures are expressed through governmental agencies which have jurisdiction over specified issues, but they are also expressed by customers and consumer groups, consumer affairs committees, and governmental consumer protection boards. Sometimes in the unfolding of the process, future policies are best expressed by the private groups responding to concerns of their members.

The utility industry is capital intensive, and all the problems of raising capital and establishing and maintaining adequate earnings must be encountered. Governmental regulations of securities' offerings, approval by regulatory commissions of these offerings, and processing them through the Securities and Exchange Commission with registration statements are matters to be weighed in implementing the process. The mechanics of the SEC are based upon the philosophy of public disclosure. As a result, the utility, responding to the public process of regulation and the disclosure requirements of the SEC, lives in a fishbowl unlike private or public entities. Not only does the utility executive have to consider all the impacts, but his considerations and decisions must be subject to the scrutiny coming from disclosure.

This issue then leads to questions of corporate control. In accordance with statutory and case law, management was always considered to be solely responsible to the shareholders. When a business enterprise was entered into for profit, management's role was clearly to develop the maximum profits as quickly as possible. All the legal mechanisms were designed to make sure that management's efforts were directed to that result; careless or inefficient management was liable to lawsuits for negligence or malfeasance. Courts have repeatedly stated that directors owe their undivided loyalty and allegiance to the company they serve.

Today, the shareholder's benefit clearly remains the objective of the corporation. Particularly in the small or private corporation, this objective remains the predominant theme. But along the way, the corporation became an instrument of other processes in addition to those involved with its shareholders. Of necessity, it became concerned with employees, customers, and the public. In a famous *Law Review* article published in 1932,

Professor Dodd of the Harvard Law School argued that corporations performed a social service as well as a profitmaking function, and that directors had a responsibility with respect to this social service.[1] Adolf A. Berle, of the New Deal and Columbia Law School, argued that directors were stewards for that part of the community affected by the corporation.[2] These discussions precipitated many debates concerning the thesis that directors were trustees for the public, but the concept remained formless and vague, and was never specifically introduced into law.

All of these discussions have been theoretical. *The utility experiment, on the other hand, has been a veritable laboratory to discover what the corporate responsibility is in daily activity.* The utility director is not only subject to corporate law, but also to the welter of regulatory laws which seek to define and impress the public interest upon the operation.

As a result, the utility experiment suggests a broader view rather than a theoretical or hypothetical one. The utility experiment, or the utility experience, indicates that management is a trustee for shareholders, employees, customers, and the public. In order to make the enterprise work, each of these groups must be satisfied. And if each is satisfied, the prosperity of all arises; none can prosper if any one of them is vigorously dissatisfied. For example, if customers lament that service is bad or rates are too high, they can make their complaints known to regulatory agencies and political organizations. If the complaints in rate cases are sufficiently vociferous, rate relief is denied or reduced. As a result, both shareholders and employees suffer. If, on the other hand, the employees do not receive adequate wages, they may perform poorly or, as a last resort, go on strike. If the shareholders are not satisfied, the company's credit may suffer and new capital becomes more difficult to obtain. Thus, management must constantly be balancing the interests of these groups, alternately making allowances and demands.

There is a point of view that public directors should be on the board of directors of a utility to represent the public interest. But these simple categories miss the point and interfere with reaching the common objective. A board must be a cohesive group of people who can work together. A harmonious board recognizing the free objectives will be far more responsive to the public interest than a devisive one. The real objective is to provide a competent nonpolitical board which will be sensitive to the public's needs. It must constantly ask, How does an issue affect the ratepayer, the shareholder, the employee, and the community?

Only the shareholders can legally make management accountable. And in this sense, management remains strictly a trustee for the shareholders. But legal pressures are not the only pressures in society; there is ample room for each of these groups to exert their demands upon the corporation and management.

Over and beyond these group influences, management must be sensitive to the demands of the community. It must listen to political inquiries. Management should also try to develop political proposals, legislative or otherwise, which are designed to assist and provide incentives to their enterprise and those who benefit from it. It must also make efforts to defeat political and legislative proposals which are harmful. Most important, it must constantly be in touch with the political and governmental agencies of its community.

Likewise, it must be aware and sympathetic to the voluntary groups which attempt to represent community interests. Consumer, business, scientific, educational, religious, and other groups must be listened to and responded to in ways consistent with the interests of all.

Thus, management in the new interacting world comes out of its parochialism and responds as a trustee—not to one special interest, but directly to the groups which comprise the corporation. In addition to responding to individuals and special interest groups, management must relate to the community as a whole.

These are the real isues of corporate control, not the narrow ones of disclosure which the SEC seeks to address in its proxy and other rules. Disclosures concerning minute details of compensation, benefits, and relationships do not go to the heart of the decisionmaking process. The real issues arise out of the pressures exerted by the competing, yet cooperative, interests involved and by the distinctive characteristics of the corporate process. And the compelling objective is to find sophisticated, highly motivated directors who will make the system work.

The utility executive must thread his way through this complicated maze. At the end of this book is a table showing the legal areas in which a utility company must seek to effect compliance. A casual review will almost assuredly call forth the response that simplification, while not necessarily the goal, is utterly essential. While the table represents the accumulative historical attempt to reflect the public interest in private enterprise in the pluralistic associative American system, it is almost certain that this worthy objective could be accomplished more efficiently.

The question then becomes not simplicity per se, but whether a mode can be adapted from the public utility experience which can produce the desired end result in an expeditious way. If regulatory procedure continues on the same basis—dominated by due process—both business and government will surely be stifled.

Perhaps a sensible procedure can be developed in which the initiatives of the private sector could be tested by a process of consultative or tentative orders of the government which narrowly define issues of fact for due process. Or they could be tested by a system of past audit to adjust after the fact, since this vain pursuit of prior determinations is thwarted by future events. Above all, there is a need for experienced, knowledgeable, objective public servants who can analyze issues in terms of the long-term public good rather than in terms of political impacts. If such a system is developed, the expertise of utility managements—which have continuously factored into all their considerations, studies, and decisions a concern for the public interest and the interest of many groups which relate to the utility process—can be instructive for the other components of our private sector system.

Such a process of interaction through consultation will bring a new vitality and a balancing of interests and considerations. These can be instrumental in preserving our system of personal freedom. The process would maintain the free system without a return to a "rugged individualism" which, even if possible, would be immediately rejected. In the same view, bland assertions about the free enterprise system do no good.

On the other hand, if we consider the attributes of the private sector in terms of the functional and social values it produces, then a line of argument emerges which has a deeply consistent relevance to basic democratic and humanitarian traditions.

Chapter 10

CONCLUSION

Public utility issues are not limited technical subjects, but ones that relate to the social issues of our day with all their over- and undertones. Each utility system is involved in the generalized issue of a national energy policy and in the specific issues of nuclear power, conservation, environment, the use of natural resources, and the pricing of electricity as they relate to local situations. All of these issues relate to the total and all-important uses of social organizations and touch on the question of the quality of life, liberty, and freedom.

The questions are: How is society best organized to give the people the maximum life satisfaction? How is society organized to express the various ways people have of judging what they want to do?

Points of view, value systems, and priorities change. Nothing can be conceived of as immutable and static. In fact, the last few years have shown a marked reversal of attitude and viewpoint.

Americans have recently expressed a social sense that life need not be progressively bigger and greater, only better and more fulfilling. The quality of life, rather than its structure, is the goal. Prestige and position are less important than the sense of the fulfillment of life. Naturalism and nature once again become prominent in people's value judgments.

Beneath the specifications of conduct, however, are qualities of life and social organization which do not change or continue to be desired. Among these qualities are freedom and liberty.

The color words "freedom" and "liberty" have been so exploited that they now sound hackneyed and outdated. Yet they still represent the glamour ideas of our history. They take on different connotations at different times and places. We romanticize them in terms of the attributes and accomplishments involved—for example, in obtaining independence from British rule, the end of slavery, and in the release from restraining customs and needless ritual. But the real meaning at any period of time lies in the functional realities of everyday life.

Because we are so accustomed to the conveniences of modern life, we hardly recognize that our independence from want depends upon the social and technical skills which modern society has developed. An oil embargo or electrical blackout quickly shows us how dependent we are on the delivery of highly technical products and services. We live on the brink of catastrophe if the nail which holds the iron shoe to the horse's hoof is lost. Indeed, modern society, which is humbled by the failure of a simple device in an interconnected electrical system, knows better than the ancients that for want of a nail, a kingdom is lost.

Thus, while it can be fun and emotionally exciting to ridicule the establishment, demonstrate against its conclusions and procedures, or urge a return to a primitive life, out of the accomplishments of established society emerge the freedom from want and drudgery which make the free life possible.

It is not so much a question of abstaining from, or doing away with, a technological society, but rather a question of molding it to human use. We need to put the two together: the mechanics of a highly specialized technical society and the aspirations of human beings who reach out for a more rewarding and satisfying life, who seek those spiritual values which bring peace, contentment, and exciting emotional experience.

So the steps must be taken to realize this putting together of the technical and the human, the mechanical and the personal. This process demands freedom and liberty and an aversion to repression. The main concern must be avoiding a political system in which human rights are submerged to the demands of the state—the one organization in human society which can deprive people of the freedom of the person and possessions. The power to put people in jail and to take away their property is an awesome one indeed—as Poland currently demonstrates—and the only way to curb that power is to strengthen the power of independent groups.

In the final result, freedom is a personal thing, but it is not achieved by individuals working on their own destinies. Instead, it depends upon free institutions and free groups. The issue, then, is whether independent and free groups can survive and prosper so that their constituents can enjoy the freedom of using initiative and developing their own ideas and personalities.

The governmental threat is an insidious thing. It comes upon us masked as the friend of the poor and the provider of wants and needs. But it results in orders and directives made by the imperfect people who are governmental officials. The task is too great, too complex, too indiscriminate to be reduced to a plan. Instead, it becomes a hodgepodge of contradictory directives which eventually, when unchallenged, lead to the frustration of unfulfilled objectives.

The process is direct when the government seeks to own and control; it is indirect and more insidious when government seeks to regulate and to set standards excessively.

At the middle road is the utility process, the interaction between the public and the private which, in effect, is a large-scale experiment in social mechanisms. It would be an overemphasis to see the utility experiment as the total answer. It would overdramatize the utility system to see it as some kind of refuge for freedom. To its critics, the thought would be ludicrous. Even its advocates would see a degree of humor in such an assertion. Yet, the utility system is clearly a social process which represents a high degree of sophistication, preserving both private incentives and the restraints of the public good.

The private initiative provides the drive and the motivation, but the governmental and community restraint on the utility injects the consciousness of the public interest. Since society no longer places its unqualified imprimatur on the bigger and the greater, and reaches instead for the quality of life, the process is the significant ingredient rather than the end result.

The question is whether human beings can remain free in a society which has become almost morbidly interdependent and precariously close to being dominated by social regulation and thought control. The answer lies in a pluralism, in a maintenance of business, educational, recreational, religious, artistic, scientific, and fraternal organizations, which have to survive and maintain their solvency. These entities then interact with each other and produce an active, vibrant competition for the energies, interests, attachment, and commitment of people.

In this interfunctionalism, the role of government becomes less a source of order and directive and more an expression of the wisdom of the community needs. The public interest becomes the protection and furtherance of the private interest as well as the individual interest—the only one which can have emotional and experiential meaning.

In all of this, the public utility, with its unique history, can provide a pattern and a guide. In this light, the public utility is entitled not only to society's acceptance, but also its active encouragement and support so the public utility can take its place with all other independent institutions in advancing the public good.

In his famous book, *The Great Bridge*, David McCullough relates the story about the mental insight which led to the construction of the Brooklyn Bridge. John A. Roebling and his son, Washington, were crossing the East River from Manhattan to Brooklyn on a ferryboat sometime in the 1860s or thereabouts, when the ferry, as it approached the Brooklyn shore, was impeded by a mass of ice floes. The boat was stranded in the river for a considerable period of time. While they were delayed, Roebling, Sr. became very impatient. Later his son, recalling the incident, stated about his father, "In his mind's eye, he saw a bridge."[1]

Subsequently, John Roebling became the designer, engineer, and constructor of the magnificent Brooklyn Bridge, which not only opened the city of Brooklyn and Long Island to the commerce of Manhattan, but also became a monument known worldwide because of the unparalleled beauty of its lines. Many years after Roebling, Judge Cardoza, writing about the processes of the law and judges, referred to the bridges around New York City, noting that these beautiful structures were a tribute to the precision and certitude which was possible in the material world. On the other hand, according to Cardoza, the processes of human institutions were inexact and unprecise.[2] In the same way as disjunctive pieces of land invite bridges, disjunctive groups and institutions invite the construction of social and institutional bridges.

In our minds's eye, we need to see those bridges that we can build to create in the modern, complex, and technological world a truly free society in which an individual can enjoy all the benefits of the modern world and still be free and creative.

Notwithstanding all our concerns for the public interest, we must remember that the only meaning of the public interest is its furthering of the private interest. In a different context, Ralph Waldo Emerson made one of his usual sage remarks which is so applicable to our concerns. In his essay entitled "Character", the following remark is found: "One day we shall discover that the most private is the most public."[3]

In short, the private should be the focus of attention. The most successful way of furthering the public good is to encourage and protect the private good. When the private flourishes, so does the public. But the reverse—the public sector—guaranteeing the success of the private—is remote and dubious.

Indeed, in this process of the continuing, growing world, the capacity of the utility company to interact with other segments of society shows us how to build those bridges between the public and private sector that will truly provide us with a design for freedom.

A PUBLIC UTILITY LEGAL COMPLIANCE OUTLINE

I. Utility Regulations
   A. Federal Department of Energy (DOE)
      1. Conservation plans under PIFUA
      2. Restrictions on the use of gas and petroleum for electric generation
   B. Federal Energy Regulatory Commission (FERC)
      1. Wholesale rates
      2. Interconnections
      3. Transmission rates
      4. Hydroelectric licensing
      5. Gas curtailment
      6. Certification of increased gas supplies and storage facilities
      7. Interlocking Positions
   C. State Public Utility Commission
      1. Rates
      2. Quality of service
      3. Financing
      4. CATV pole attachments
      5. Home insulation and energy conservation programs
      6. Filing of cost-plus contracts
   D. Siting Boards
      1. For location of generating plants
      2. For location of transmission lines
   E. Long range plans (State Energy Office)
      1. Electric and gas long range forecasts
      2. Research and development programs

II. Nuclear Regulatory Agency

III. Environmental Regulation
   A. Environmental Protection Agency (EPA)
      1. Clean water
         a. EPA effluent guidelines
         b. Thermal pollution
         c. National Pollutant Discharge Elimination System permits
      2. Clean air
         a. Regulations under Clean Air Act
         b. Requirements regarding existing sources and new and modified sources

3. Hazardous and toxic substances
   a. Regulations under RCRA for active sites
   b. Regulations under CERCLA for inactive sites
   c. Notification of releases to environment
B. State Environmental Department
  1. Clean water
  2. Clean air
  3. Hazardous and toxic substances
C. Environmental Impact Statements
  1. Federally authorized activities
  2. State authorized activities

IV. Regulation of Employee Relations
  A. Employee
    1. Union organizing and settlement of labor disputes
    2. Wage and hours
    3. Strikes
    4. Workmen's Compensation
  B. Pension Regulations
    1. ERISA and related regulations
    2. Benefit plan compliance with applicable rules
    3. IRS regulations
    4. Department of Labor
    5. Pension Benefit Guaranty Corporation
  C. Discrimination in Employment
    1. Equal Employment Opportunity Compliance
       Guidelines and Enforcement
    2. Federal Contract Compliance
       OFCC regulation and enforcement
       General Services Administration regulations

V. Occupational Safety and Health Administration (OSHA)
  A. Standards Compliance
  B. Standards Enforcement

VI. Antitrust and Trade Regulation
  A. Justice Department
  B. Federal Trade Commission
  C. Private Antitrust Actions

VII. Taxation
  A. Federal
    1. Income tax
      a. Enforcement, Audits, Appeals

2. Excise Taxes
    a. Highways
    b. Fuels
    c. Communities
3. Social Security and Medicare
4. Withholding
    a. Payroll
    b. Pension
    c. Dividends
    d. Interest
5. Payroll taxes—unemployment
6. Reporting
B. State and Local
1. Property
2. Sales
3. Income
4. Franchise and special franchise

VIII. Securities and Exchange Commission
A. Financing
    1. Registration
    2. Exemptions
B. Stockholder Relations
    1. Annual reports
    2. Proxy and proxy statements
C. Reporting
    1. Annual reports
    2. Quarterly and current reports
D. Stock Exchanges
    1. Listing requirements
E. Corporate Governance
    1. Audit committees
    2. Foreign Corrupt Practices Act
F. Holding Company Act Regulations

IX. Corporate
A. Political Contribution
B. Federal Election Laws
C. State Election Laws
D. State Requirements Regarding Corporate Charters and Amendments
E. Lobbying Laws
F. State Blue Sky Laws

X. Debt Instruments
    A. Mortgage Indenture
    B. Unsecured Debt Agreements
    C. Other

XI. Credit Regulation
    A. Consumer Credit Protection Act
    B. Federal Reserve Board Requirements

XII. Miscellaneous
    A. Federal Contract Rules
    B. Patents and Trademarks
    C. Federal and State Reporting Requirements

NOTES

Chapter 1

1. Paul Freund, "Two Cheers for the Court," 19 *Cleveland State Law Review* 7, 1970.

2. William Wordsworth, "Lines Composed a Few Miles Above Tintern Abbey."

3. William James, *The Moral Philosopher and the Moral Life.*

4. The author attended the hearing.

5. Learned Hand, *The Spirit of Liberty*, 3rd ed. enl. (Chicago: University of Chicago Press, 1960), pp. 229–230.

6. Ibid.

Chapter 2

1. Daniel J. Boorstin, *The American: The Colonial Experience* (New York: Vintage Books, 1958), p. 175.

Chapter 3

1. Martin Buber, *I and Thou*, 2nd ed., trans. Ronald Gregor Smith (New York: Charles Scribner, 1958), p. 11.

2. *Sutton's Hospital* Case, 10 Coke 23 (1613).

3. Dartmouth College Case, 4 Wheat 518, 4 L. Ed. 629.

4. Ibid.

5. Walter Lippman, quoted in Central Hudson Gas & Electric Corporation's 1974 annual report.

Chapter 4

1. John Donne, "Death Be Not Proud."

2. James Kent, *Commentaries on American Law*, vol. 2, 5th ed. (New York: William Kent, 1854), p. 346.

3. Max Weber, *The Theory of Social and Economic Organization* (London: Oxford University Press, 1947), p. 337.

4. Bruce L. R. Smith, ed., *The New Political Economy: The Public Use of the Public Sector* (New York: Halstead Press, 1975).

Chapter 5

1. *Electric Perspectives*, Summer 1982 (Washington D.C.: Edison Electric Institute), p. 45.

Chapter 9

1. Professor Dodd, "For Whom Are Corporate Managers Trustees," 45 *Harvard Law Review* 1049, 1931.

Chapter 10

1. David McCullough, *The Great Bridge,* rev. ed. (New York: Avon, 1976).

2. Benjamin Cardoza, *Selected Writings,* ed. Margaret E. Hall (San Jose, Cal.: James R. Bender Publishing, 1947).

3. Ralph Waldo Emerson, "Character."

ABOUT EEI

Edison Electric Institute is the association of America's investor-owned electric utility companies. Organized in 1933 and incorporated in 1970, EEI provides a principal forum where electric utility people exchange information on developments in their business, and maintain liaison between the industry and the federal government. Its officers act as spokesmen for investor-owned electric utility companies on subjects of national interest.

Since 1933, EEI has been a strong, continuous stimulant to the art of making electricity. A basic objective is the "advancement in the public service of the art of producing, transmitting, and distributing electricity and the promotion of scientific research in such field." EEI ascertains factual information, data, and statistics relating to the electric industry, and makes them available to member companies, the public, and government representatives.